THE COMSOC GUIDE TO PASSIVE OPTICAL NETWORKS

A volume in the IEEE Communications Society series:
The ComSoc Guides to Communications Technologies

Nim K. Cheung, *Series Editor*
Thomas Banwell, *Associate Editor*
Richard Lau, *Associate Editor*

Next Generation Optical Transport: SDH/SONET/OTN
Huub van Helvoort

Managing Telecommunications Projects
Celia Desmond

WiMAX Technology and Network Evolution
Edited by Kamran Etemad, Ming-Yee Lai

*An Introduction to Network Modeling and Simulation for
the Practicing Engineer*
Jack Burbank, William Kasch, Jon Ward

THE COMSOC GUIDE TO PASSIVE OPTICAL NETWORKS
Enhancing the Last Mile Access

STEPHEN WEINSTEIN
YUANQIU LUO
TING WANG

IEEE
COMMUNICATIONS
SOCIETY

The ComSoc Guides to Communications Technologies
Nim K. Cheung, *Series Editor*
Thomas Banwell, *Associate Series Editor*
Richard Lau, *Associate Series Editor*

IEEE

IEEE PRESS

WILEY

A JOHN WILEY & SONS, INC., PUBLICATION

Published by John Wiley & Sons, Inc., Hoboken, New Jersey.
Published simultaneously in Canada.

For general information on our other products and services or for technical support, please contact our Customer Care Department within the United States at (800) 762-2974, outside the United States at (317) 572-3993 or fax (317) 572-4002.

Wiley also publishes its books in a variety of electronic formats. Some content that appears in print may not be available in electronic formats. For more information about Wiley products, visit our web site at www.wiley.com.

Library of Congress Cataloging-in-Publication Data:

Weinstein, Stephen B.
    The ComSoc guide to passive optical networks : enhancing the last mile access / Stephen B. Weinstein, Yuanqiu Luo, Ting Wang.
        p. cm.
    ISBN 978-0-470-16884-4 (pbk.)
    1. Passive optical networks. I. Luo, Yuanqiu. II. Wang, Ting. III. Title. IV. Title: Guide to psssive optical networks.
    TK5103.592.P38W45 2012
    621.382'7–dc23
                                        2011037610

10 9 8 7 6 5 4 3 2 1

To my wife, Judith
Stephen Weinstein

To my family
Yuanqiu Luo

To my children
Ting Wang

CONTENTS

3 Techniques and Standards   **53**

4 Recent Advances and Looking to the Future   **87**

PREFACE

This handbook is a convenient reference guide to the rapidly developing family of passive optical network (PON) systems, techniques, and devices. Our objective is to provide a quick, intuitive introduction to these technologies, with clear definitions of terms, including many acronyms. We have avoided extensive technical analysis.

PON provides a high ratio of performance to cost for high-speed data network access, making possible an economical successor to DS-1 and DS-3 services and promising stiff competition for alternative access technologies such as cable data in hybrid fiber/coax (HFC) systems, digital subscriber line (DSL), broadband over power line, and broadband wireless. At the same time, PON provides attractive opportunities for integration with other access systems and technologies and, in particular, for integration with very high-speed DSL and with broadband wireless access systems. The goals are enhancement of overall capacity, reliability, and peak-load performance at minimum cost. This book will describe both the competitive and the cooperative potential of PON technologies.

As a well-indexed reference work, this book should provide quick answers to questions about PON terminology, definitions, and basic operational concepts while encouraging the reader to acquire a deeper understanding of PON capabilities and of the entire broadband access environment. PON already has a very important role in realizing per-user access rates in the hundreds of megabits per second and an access infrastructure that truly serves the needs of a global information society.

STEPHEN WEINSTEIN
YUANQIU LUO
TING WANG

1

PON IN THE ACCESS PICTURE

1.1 WHY PASSIVE OPTICAL NETWORK (PON) FOR THE LAST MILE ACCESS?

As part of the telecommunications network, the access network covers the "last mile" of communications infrastructure that connects individual subscribers to a service provider's switching or routing center, for example, a telephone company's central office (CO). We will use CO, a term from the traditional public network, for convenience, although the switching or routing center could be operated by any entity under a different name, such as headend. The access network is the final leg of transmission connectivity between the customer premise and the core network. For a variety of access solutions including the PON, the access network consists of terminating equipment in the CO, a remote node (RN), and a subscriber-side network interface unit (NIU), as Figure 1.1 shows. The feeder network refers to the connection between CO and RN, while the distribution network joins the NIU to the RN. Downstream program services, one of many applications of a broadband access system, may be broadcast, multicast, or individually directed to the users, depending on the service objectives and enabling technologies.

The access network has consistently been regarded as a bottleneck in the telecommunications infrastructure [GREEN]. This is primarily because of the ever-growing demand for higher bandwidth, which is already available in large

The ComSoc Guide to Passive Optical Networks: Enhancing the Last Mile Access,
First Edition. Stephen Weinstein, Yuanqiu Luo, Ting Wang.
© 2012 Institute of Electrical and Electronics Engineers. Published 2012 by John Wiley & Sons, Inc.

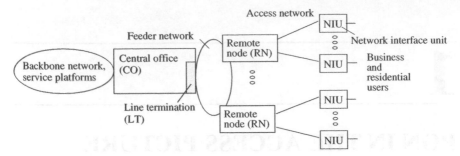

Figure 1.1 Generic access network architecture.

measure in the core optical network and in local area networks (LANs) but is more limited in widely deployed residential access technologies such as digital subscriber line (DSL) and cable data. Business customers using the relatively expensive DS-1 (1.544 Mbps) and DS-3 (45 Mbps) legacy access services are similarly limited. We have, then, a large disparity between legacy access systems with per-user rates in the low megabits per second, and the network operator's optical backbone network using multiple carrier wavelengths in wavelength division multiplexing (WDM) systems in which each wavelength carries data at rates of tens of gigabits per second. The disparity between legacy access systems and both wired and wireless LANs, which have been scaled up from 10 to 100 Mbps and are being upgraded to gigabit rates, is equally dramatic. The tremendous growth of Internet traffic accentuated the growing gap between the capacities of backbone and local networks on the one hand and the bottleneck imposed by the lower capacities of legacy access networks in between. This was, and in many cases still is, the so-called last mile or last kilometer problem. Upgrading the current access network with a low-cost and high-bandwidth solution is a must for future broadband access, and is being actively implemented by many operators.

Operators expect that large capacity increases in the access network, facilitated by advances in enabling technologies, will stimulate diverse services to the customer premise and new revenue streams. To realize truly high-speed broadband access, major worldwide access providers, including, but not limited to, AT&T, Verizon, British Telecommunications (BT), and Nippon Telegraph and Telephone (NTT), are making significant investments in fiber-to-the-home (FTTH) and broadband wireless access (BWA). Among the many possible wired approaches, the PON (Figure 1.2) is especially attractive for its capability to carry gigabit-rate network traffic in a cost-effective way [LAM]. In comparison with very high-speed digital subscriber line (VDSL) and cable data infrastructure, which requires active (powered) components in the distribution network, PONs lower the cost of network deployment and maintenance by employing passive (not powered) components in the RN between the optical line terminal (OLT) and optical network unit (ONU) or terminal (ONT).

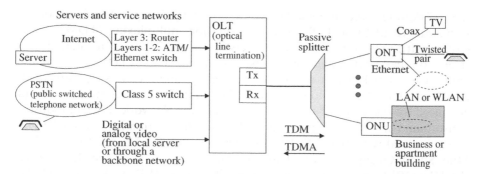

Figure 1.2 Generic PON, shown delivering "triple play" services (Section 1.2.3).

A decision for deployment of PON depends, of course, on the operator's perception of revenues versus costs. Investment must be made in the following [BREUER]:

- the aggregation link in the backhaul network between a PON access site, where the OLT, possibly heading several PONs, is located (shown as a CO in Figure 1.1), and a transport network point of presence;
- the PON access site itself, where the RN is located;
- the feeder links between the OLT and the passive splitters of the several PONs; and
- the "first mile" including a passive splitter and its access lines to user optical network terminations (ONTs or ONUs).

As access sites are more densely deployed, the total per-ONT cost initially decreases due to shorter links through the feeder network. However, beyond a certain optimum density of access sites, cost climbs as the costs of access sites and aggregation links begin to overwhelm the savings from shorter feeder links. As noted in [BREUER], with appropriate selection of access sites, PON is significantly less expensive than active optical fiber access networks, perhaps by a factor of two in relation to point-to-point (P2P) gigabit Ethernet, which is not only more expensive but also consumes much more energy.

Note that the OLT corresponds to the line termination (LT) in Figure 1.1, the splitter to the RN in Figure 1.1, and the ONU to the NIU in Figure 1.1. The terms ONU and ONT are sometimes used interchangeably, although the ONU may have additional optical networking connected to its subscriber side, while the ONT does not.

The PON standards of current interest include broadband passive optical network (BPON) [ITU-T G.983.1], Ethernet passive optical network (EPON) (Institute of Electrical and Electronics Engineers [IEEE] 802.3ah incorporated into IEE 802.3-2008), gigabit-capable passive optical network (GPON) [ITU-T G984.1], and 10G PON (IEEE 802.3av-2009 and ITU-T G.987). Note that "x" denotes several possible integers denoting different documents of the standard. All of these PONs use time division multiplexing (TDM) downstream, with data sent to different users in assigned slots on a single downstream optical carrier, and time division multiple access (TDMA) upstream, with greater flexibility in requesting and using time on the single upstream optical carrier. The development of BPON and GPON was stimulated and advanced by the work of the Full Service Access Network (FSAN) industry consortium (http://www.fsanweb.org). The standardization process for EPON began in 2000 with IEEE 802.3's establishment of the Ethernet in the First Mile Study Group and the later formation of the P802.3ah Task Force [ECHKLOP]. Work is also in progress on WDM-PON [BPCSYKKM, MAIER] for future large increases in capacity following from the use of multiple wavelengths.

This guidebook covers the major concepts and techniques of PONs, including components, topology, architecture, management, standards, and business models. The rest of this chapter introduces nonoptical access technologies and important features of the entire family of optical access systems, which we collectively denote as fiber-to-the-building (FTTB), fiber-to-the-business, fiber-to-the-cabinet (FTTCab), fiber-to-the-curb (FTTC), FTTH, fiber-to-the-node, fiber-to-the-office, fiber-to-the-premise, and so on, or FTTx. Section 1.4.1 offers additional defining information about PON. Chapter 2 covers PON architecture and components, elaborating on the major alternatives introduced above. Chapter 3 describes PON techniques and standards, largely in the physical level (PHY) and medium access control (MAC) layers of the protocol stack. Chapter 4 describes recent advances, particularly WDM-PON, interoperability with other optical networks, and what is coming in the near future, including wireless/optical integration.

1.2 SERVICES AND APPLICATIONS

PONs offer many possibilities for service replacement and for support of applications, in both residential and business markets. We describe here several of the most significant, beginning with replacement of other high-speed access services such as asymmetric digital subscriber line (ADSL), very high speed digital subscriber line (VDSL), cable data, DS-1, and DS-3.

1.2.1 Displacement of Legacy High-Speed Access Services

The nonoptical, copper-based "broadband" access services offer downstream burst data rates ranging from hundreds of kilobits per second to about 10 Mbps, and many of them are asymmetric with considerably lower upstream data rates. The average data rate per subscriber may be further limited to something well below the maximum burst rates. Much higher rates are possible under recent standards but are not commonly deployed. Several of these services will be described in the next section, together with the still developing broadband over power line (BoPL) and BWA, including mesh IEEE 802.11 (Wi-Fi) and IEEE 802.16 (Worldwide Interoperability for Microwave Access [WiMAX]) networks.

For mobile users and applications, PON cannot replace the wireless alternatives. It can, in fact, enhance them, as discussed in Chapter 4. But for fixed residential users, PON can yield a higher ratio of performance to cost than any of the available alternatives that are described in Section 1.3. Its current data rates of 50–100 Mbps per subscriber in both downstream and upstream directions compare favorably with the commonly deployed version of the fastest copper-based system, VDSL, with its 50 Mbps divided between downstream and upstream traffic. Of equal importance is the fact that the RN of a VDSL system is active, unlike the passive RN of a PON system, requiring more initial outlay and recurrent maintenance expense. The cost and other advantages of PON, over various active access systems, as noted in [SHUMATE], include its elimination of

- active optoelectronic and electronic devices operating in an often harsh outside environment,
- power conversion equipment and backup batteries in that location,
- electromagnetic interference (EMI) and electromagnetic compatibility (EMC) issues,
- energy costs, and
- environmental controls.

In addition, a PON node reduces the failure rate and associated repair costs typical of powered nodes, and its bandwidth-independent components allow future upgrades at minimal cost.

For a "greenfield" deployment without existing wiring, the total initial investment is comparable for both VDSL and PON, and the PON advantage in capability and lower maintenance cost is clear. For a PON overbuild, on top of an existing copper plant, the initial investment for PON is greater than that for VDSL because of the added expense of deploying the new optical distribution network. It is difficult to claim a clear economic advantage for PON in this case, but there is a compelling case in its higher current data rate and the possibility of much higher rates through future deployment of WDM end equipment without modifying the passive splitter.

For business customers, PON can provide higher data rates at costs lower than those of DS-1 and DS-3 services. Network operators are motivated to make this replacement because of the lower maintenance costs and much greater service flexibility, allowing easy changes in capacity allocations to different users served by the same PON splitter. PON services that are currently available, mostly BPON and GPON in the United States and EPON (1 Gbps in each direction) in East Asia, are offered at costs that are very competitive in comparison to legacy DS-1 (1.5 Mbps) and DS-3 (45 Mbps). Even when shared among a number of users, GPON's data rates (2.5 Gbps downstream and 1.5 Gbps upstream with 10 Gbps downstream being introduced) and EPON's data rates (1 Gbps in each direction with 10 Gbps being introduced) compare favorably with those of the legacy services.

1.2.2 Internet Protocol (IP) over PON

All of the current and contemplated access systems support IP traffic to a lesser or greater extent. BPON is oriented toward circuit-switched traffic, either asynchronous transfer mode (ATM) or TDMA, but the newer GPON and EPON are designed to transport variable-sized packets such as IP traffic, which is terminated in a packet router in the CO.

EPON, in particular, utilizes a flexible multipoint control protocol (MPCP) defined in IEEE 802.3ah to coordinate the upstream transmissions of different users. This protocol supports dynamic bandwidth allocation (DBA) algorithms [AYDA] that, in turn, support the Internet's differentiated services [WEIN] for heterogeneous traffic including voice over Internet protocol (VoIP) and Internet protocol television (IPTV). Because of capabilities like this one, in addition to low-cost bandwidth, PON access systems are likely to accelerate the transition to converged applications based on IP, as suggested in the next section, particularly in Figure 1.3.

1.2.3 Triple Play and Quadruple Play

"Triple play" is a package of video, voice, and high-speed Internet services on a single access system, and "quadruple play" extends the package to include wireless services. The profitability of these packages is one of the main motivators of carriers to pursue the deployment of FTTx in the broadband access network. TDM-PON technologies such as BPON, GPON, and EPON are widely adopted to enable the delivery of triple play to subscribers, as shown in Figure 1.2 for the configuration used in current deployments. Figure 1.3

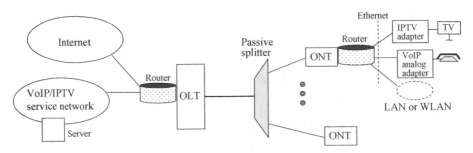

Figure 1.3 Triple play over an IP-oriented PON, with illustrative residential networking.

TABLE 1.1 Triple and Quadruple Play Services

Category	Services	Category	Services
Data	High-speed Internet	Wireless	Wi-Fi
	Private lines		WiMAX
	Frame relay		Cellular pico/femtocells
	ATM		Ultra-wideband (UWB)
Voice	Plain old telephone service (POTS)		Medium-speed Internet
	VoIP		Multimedia "apps"
Video	Digital broadcast video		
	Analog broadcast video		
	High-definition television (HDTV)		
	Video on demand (VoD)		
	Interactive TV		
	TV pay per view		
	Video blog		

shows the configuration likely in the future in which voice will be VoIP and video will be IPTV, all fed into the Internet or IP-based networks dedicated to higher-quality services. Table 1.1 tabulates the triple and quadruple play services available over PONs.

PONs provide several advantages to operators for the delivery of triple play services. There is plenty of bandwidth for all three services. Tens of subscribers share one feeder fiber, minimizing field costs, and also share wavelengths and transmission equipment at the CO. The use of a single access system for all services minimizes operations and maintenance costs.

At the CO, in the current implementation (Figure 1.2), Internet and public switched telephone network (PSTN) services enter the PON access system via an IP router and a class 5 switch, respectively. Diverse video signals are converted to an optical format in the optical video transmitter. The OLT aggregates various services and distributes them through the PON. At the subscriber side, existing twisted-pair cable may be employed to deliver the telephone service, while10/100 Base-T Ethernet cable and Wi-Fi wireless LAN are often used for data service delivery. The video broadcast service is transmitted through a coax cable to the set-top box (STB) and then to the TV set. In the all-IP future system of Figure 1.3, a wide variety of in-home local networks may be used, including Ethernet, IEEE 1394 Firewire, ultra-wideband (UWB), IEEE 802.11 Wi-Fi, IEEE 802.16 WiMAX, and power line communication (PLC) systems.

The triple play architecture employs different devices for different services and requires a delicate balance among the various demands. Video service typically requires high bandwidth, medium latency (transmission and processing delay), and very low loss. Data may require medium bandwidth with variable latency and either low or moderate losses. Voice consumes less bandwidth than data or video and tolerates low loss but requires very low latency. As a result, the TDM-PON pipe needs to be designed to provide the following features:

- Bundled services (a set of several different services sold as one package)
- A service level agreement (SLA) specifying access requirements and limits for each service
- Quality-of-service (QoS) mechanisms
- Differentiated services with different traffic treatments

DBA in the upstream direction and QoS provisioning downstream, both discussed later in this book, are the critical approaches to obtaining these features.

1.2.4 Multimedia Conferencing and Shared Environments

Multimedia conferencing is a lot like triple play, combining different media elements, but has special requirements for synchronizing these elements into a single presentation. PON access systems, again, provide adequate capacity for this multimedia application, which can consume tens of megabits per second or more for current and future high-definition applications. Moreover, through coordinated traffic scheduling in the access network, they can assume some of the burden of differentiated QoS and synchronization of media streams. This combination of high-capacity and flexible-capacity scheduling is a powerful incentive for the innovation of new conferencing systems and applications including online panel discussions with audiovisual response from the audience; large-screen, high-definition "video window" systems; and elaborate three-dimensional immersive environments combining computer-generated elements with human participation. This last category includes sharing real-time games in realistic environments and multiperson training in simulated dangerous environments. Removing the access bottleneck can and will release a new burst of creative design of shared applications and experiences.

1.2.5 Backhaul Services

Backhaul refers to connection of remote traffic aggregation points to the metropolitan backbone network. Traffic aggregation points include business and residential wired LANs (e.g., Ethernets), wireless access points in municipal or home Wi-Fi or WiMAX networks, and cellular mobile base stations (BSs) or BS control points. Figure 1.4 illustrates these possibilities. At present, most of the mobile operators lease T1/E1 copper lines to bridge mobile networks to the core infrastructure. The proliferation of high-speed wireless appli-

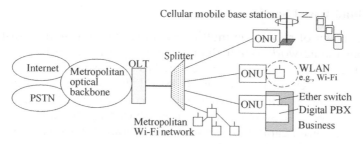

Figure 1.4 PON backhaul applications.

cations will create a bottleneck in mobile operators' backhaul links, with the current T1/E1 copper lines unable to provide the required capacity. Increased demand for BWA will likely lead to a proliferation of "femtocells," very small mobile cells for high-capacity media traffic in businesses, apartment buildings, and public places, creating the need for a greatly extended backhaul system. This is an aspect of optical–wireless integration discussed in Chapter 4. PON would probably find it easier than alternative access networks that use the IEEE 1588 synchronization protocol to implement the timing needed for smooth call handoff.

The PON architecture satisfies the requirement of high-speed backhaul from varied access traffic aggregation points. Figure 1.4 illustrates numerous neighborhood ONUs, which, as we noted earlier, may be called ONTs if there is no additional optical networking between them and end users. Depending on the split ratio and PON capacity, each ONU/ONT might support aggregated traffic ranging from several megabits per second up to tens or even hundreds of megabits per second, with low latency. The advantages of PONs for backhaul include

- Larger capacity than a leased T1/E1 line while retaining excellent timing synchronization capabilities
- On-demand bandwidth flexibility
- Scalability as network requirements grow

Over the long term, WDM-PON, with its very high capacity and support of disparate data formats and data rates from its associated ONUs/ONTs, promises to become the preferred backhaul solution among the different PON technologies. As technologies traditionally associated with wireless communication, particularly orthogonal frequency division multiplexing (OFDM), are exploited by optical communications engineers, we can also expect development of systems such as OFDM-PON that offer services and management flexibilities including transport of broadband wireless signals [OFDMPON].

1.2.6   Cloud-Based Services

The old concept of relying on multiuser computational resources distributed throughout the network, rather than maintaining private resources such as a dedicated corporate database or server, has been somewhat redefined in recent years as "cloud computing" [CISCO]. In essence, cloud computing facilitates the provisioning of virtual resources, abstracted from the actual underlying physical resources, that can be shared by multiple users in the interests of reduced capital investment and operating costs, greater capacity, and more powerful capabilities. There are concerns about security and privacy and about vulnerability to limitations on access or performance caused by adverse network conditions, whether natural or malicious.

High-capacity access with flexibility attributes is an obvious need for successful delivery of cloud-based services. PON thus helps meet an important prerequisite for further replacement of dedicated facilities by distributed virtual resources. This is true not only for enterprises seeking to reduce dependence on expensive private "back-office" servers and databases but also for offering much greater service capabilities to consumers and other end users. One example might be making "apps" (applications for smart phones and other personal devices) available in multiple versions for different devices, realizing application transportability that has been difficult to realize by more conventional methods.

1.3   LEGACY ACCESS TECHNOLOGIES

The competition for PON is the range of currently deployed and still developing legacy access technologies. PON could develop as a natural extension or upgrade of some of these legacy systems.

1.3.1   Hybrid Fiber-Coax (HFC) Cable Data System

The traditional analog cable television (CATV) system supports tens of downstream TV channels for news, entertainment, and educational programs. Each analog TV channel occupies a 6-MHz slot (in North America) or an 8-MHz slot (in Europe) in the cable's available frequency band.

A cable data system is an extension of the original CATV concept with digital signals, both downstream video programming and interactive data. This was made possible, in large part, by a massive upgrading from all coaxial cable to HFC systems that cable operators made some time ago, more to enhance reliability and to reduce maintenance costs than to support new digital services. In HFC systems, a high-speed digital signal, typically conveying data at 30 Mbps, replaces an analog video signal in each downstream 6- or 8-MHz slot. The high spectral efficiency comes from the use of multilevel modulation formats, such as 64-point quadrature amplitude modulation (QAM), in rela-

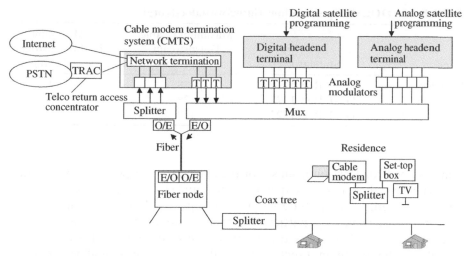

Figure 1.5 HFC network for cable data and video services.

tively good transmission channels. Six or seven MPEG-2 digital television signals, or one digital high-definition television (HDTV) signal plus one or two ordinary digital television signals, occupy the bandwidth formerly needed by just one analog video signal, a tremendous benefit for the operator. Alternatively, a 30-Mbps downstream signal may provide Internet traffic to dozens of Internet access subscribers. In the upstream direction from a user to the Internet, a TDMA system multiplexes the user's traffic with that of other users into narrower (typically 1.5 MHz) channels, reflecting the assumption that people download far more information than they upload, which may not be true indefinitely. Figure 1.5 illustrates an HFC system supporting both analog and digital services. Like PON, it uses a two-tier architecture with an RN, called the fiber node, but unlike PON, the fiber node is active, not passive, and does optical-electrical conversions.

A cable data system requires a cable modem on the user end and a cable modem termination system (CMTS) at the cable provider's end [FJ]. As shown in Figure 1.5, in the customer premises, a one-to-two splitter provides a coaxial cable line to the cable modem and another coaxial line to a TV STB or the TV itself. The cable modem connects, in turn, to a router, a wireless router, or a computer through a standard 10 Base-T Ethernet or Universal Serial Bus (USB) interface. Appropriate modulators and filters separate the upstream and downstream signals into their respective disjoint lower and upper ranges.

The CMTS located at the cable operator's network headend is a data switching system that routes data to and from many cable modems over a multiplexed network interface. For downstream traffic, the cable headend uses different channels in the HFC network for data, video, and audio traffic and broadcasts them throughout the HFC network except, that is, for "on-demand" programs

TABLE 1.2   DOCSIS Specifications (http://www.docsis.org)

Version	Standard	Maximum Usable Downstream Speed (Mbps)	Maximum Usable Upstream Speed (Mbps)
DOCSIS 1.1	ITU-T Rec. J.112	38	9
DOCSIS 2.0	ITU-T Rec. J.122	38	27
DOCSIS 3.0	ITU-T Rec. J.222	152	108

that may be broadcast on a subset of the network corresponding to the locations of active customers. The downstream digital channels use QAM as noted above, a system for amplitude modulating separate data pulses on both the sine and cosine waves at a particular carrier frequency, supporting a total data rate of $2 \log_2 N$, where N is the number of possible levels of a data pulse [GHW]. In the upstream direction, the multiple access system allocates "minislots" to different users according to demand. Traffic is routed from the CMTS to the backbone of a cable Internet service provider (ISP) or, alternatively, for telephone service, to the PSTN after appropriate protocol conversions.

Cable data systems follow the Data over Cable Service Interface Specification (DOCSIS) drafted by CableLabs (http://www.cablelabs.com), an industry-supported institution, to promote cable modem rollouts in 1996. Table 1.2 lists the main specifications that are available from the CableLabs Web site. DOCSIS 2.0 (2001) improves on DOCSIS 1.1 (1999) by substantially increasing upstream channel capacity, using denser QAM modulation with greater spectral efficiency and enhancing error correction and channel equalization. DOCSIS 3.0 (2006) improves on DOCSIS 2.0 by "channel bonding" to increase both downstream and upstream peak burst rates, enhancing network security, expanding the addressability of network elements, and deploying new services offerings. The International Telecommunication Union-Telecommunications (ITU-T) standardization sector adopted three DOCSIS versions as international standards. DOCSIS 1.0 was ratified in 1998 as ITU-T Recommendation J.112; DOCSIS 2.0 was ratified as ITU-T Recommendation J.122; and DOCSIS 3.0 was ratified as ITU-T Recommendation J.222 (http://www.itu.int/itu-t/recommendations/index.aspx?ser=J).

Cable modem users in an entire neighborhood share the available bandwidth provided by a single coaxial cable line through time slot allocations. Therefore, connection speed varies depending on how many people use the service and to what extent users simultaneously generate traffic [DR]. Of course, as in every access system, new capital investment can increase capacity, in this case, by setting up additional fiber nodes lower in the distribution tree. While cable modem technology can theoretically support 30 Mbps or more, most providers offer service with data rates between 1 and 6 Mbps downstream, and between 128 and 768 Kbps upstream. In addition to the signal fading and crosstalk problems introduced by the coaxial cable line, cable modem service has additional technical difficulties including maintenance of the active fiber nodes. Also, increasing upstream transmission capacity inevitably encroaches on the downstream capacity (in the cable part of the plant), which may not be to the liking of the video services provider [AZZAM]. These are weaknesses in comparison with PON, but there is a lot to say for "being there" with an effective system for both video program distribution and interactive data communication. As of September 2011, 130 million homes were passed in the United States, almost equal to the total number of homes, more than 77% of which had access to HDTV services and 93% to high speed Internet service (http://www.ncta.com/Statistics.aspx). Of the homes passed, 45% subscribe to at least basic cable video services.

1.3.2 Digital Subscriber Line (DSL)

DSL is a family of technologies for digital data transmission over the twisted-pair copper subscriber line of a local telephone network. Although described here as a legacy technology, recent enhancements, including short-range 100-Mbps VDSL relying on an optical RN just as cable HFC does, and vectored transmission [GC] for crosstalk cancellation in bundled pairs, enable high performance comparable to current-generation PON.

The twisted-pair subscriber line between a subscriber and a telephone office or RN is the same wiring used for "plain old telephone service" (POTS), which occupies only a 300- to 3300-kHz portion of what is actually a much wider (around 1 MHz) usable bandwidth. Voiceband modems, sending data through the voice network, are constrained by voice filters in the telephone office to this small band. The demand of more bandwidth has resulted in exploiting the remaining capacity on the subscriber line to carry data signals without, in some cases, disturbing its ability to carry voice services. Table 1.3 illustrates some of the alternative DSL formats. Of these, only ADSL supports simultaneous POTS.

In particular, ADSL carries voice signals in the usual 300- to 3300-Hz voiceband and two-way data signals on the unused higher frequencies [WEIN], allowing simultaneous telephony and Internet access. The downstream data

TABLE 1.3   DSL Technologies

xDSL	Standard	Downstream	Upstream	Symmetry
ADSL	ITU-T Rec. G.992.1	Up to 8 Mbps	Up to 1 Mbps	Asymmetric
HDSL	ITU-T Rec. G.991.1	784 Kbps, 1.544 Mbps, 2.0 Mbps	784 Kbps, 1.544 Mbps, 2.0 Mbps	Symmetric
SDSL	–	Up to 2 Mbps	Up to 2 Mbps	Symmetric
VDSL	ITU-T Rec. G.993.1 ITU-T Rec. G.993.2	Up to 100 Mbps	Up to 100 Mbps	Both

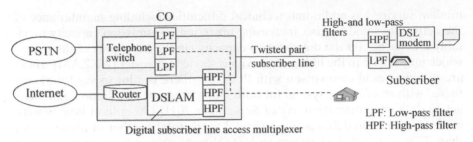

Figure 1.6 DSL network.

signal uses discrete multitone (DMT) transmission, in which the bandwidth is segmented into a large number—typically 256—frequency division multiplexed channels, each about 4 kHz wide. Fast Fourier transform (FFT) enables an efficient computational algorithm for generating these parallel channels. DMT is essentially the same as OFDM cited in Section 1.3.4.

As shown in Figure 1.6, DSL service is distributed through the P2P dedicated public network access between a service provider CO and a user. DSL modems in the customer premises contain an internal signal splitter. It separates the line serving the computer from the line that serves the POTS devices, such as telephones and fax machines located at a telephone company.

CO, digital subscriber line access multiplexers (DSLAMs) receive signals from multiple DSL users. Each DSLAM has multiple aggregation cards, and each such card can have multiple ports to which the DSL lines are connected. The DSLAM aggregates the received signals on a high-speed backbone line using multiplexing techniques. DSL appeals to telecommunications operators because it delivers data services to dispersed locations using already-installed copper wires, with relatively small changes to the existing infrastructure.

The DSL family covers a number of similar yet competing forms of DSL technologies, including ADSL, symmetric digital subscriber line (SDSL), high-bit-rate digital subscriber line (HDSL), rate-adaptive digital subscriber line (RADSL), and very high speed DSL, as described in Table 1.3 [SSCP]. DSL is distance sensitive, and the supported data rate varies depending on the transmission length. Essentially, customers with longer telephone line runs from their houses to the CO experience lower performance rates as compared to neighbors who might live closer to the CO [ODMP]. For the conventional ADSL, downstream rates start at128 Kbps and typically reach 8 Mbps in the wire length of 1.5 km; upstream rates start at 64 Kbps and can go as high as 1 Mbps within the same distance. This dependence on distance is a weakness of DSL, especially in the United States, where there are many long subscriber lines and as many as 20% of telephone subscribers cannot be served by ADSL at a desirable data rate. Of course, with capital investment in RNs, the copper wire runs are drastically shortened and high VDSL data rates become possible.

The major problems with sending a high-frequency signal, such as DSL, over an unshielded pair of copper wires include signal fading and cross talk. As the length of wires increases, the signal at the customer side may become too weak to be correctly detected even with aggressive channel equalization. If downstream transmitted power is increased at the CO, the signals tend to transfer to other subscriber lines in the same bundle. Cross talk, especially near-end cross talk (NEXT) from local transmitters, severely impairs service, although there are vectored techniques, noted earlier in this subsection, that can significantly improve performance. "100 Gbps DSL Networks" might possibly offer serious competition to PON [CJMG].

1.3.3 Broadband over Powerline (BoPL)

BoPL uses PLC technology for broadband communication services through the electrical power supply networks. Power line communications carries modulated carrier waves on the power line together with the usual 50- to 60-Hz electric current, and simple filters easily separate them. Special bypasses are needed around transformers, which would otherwise greatly attenuate the high-frequency information signals. By slightly modifying the current power grids with specialized equipment, power companies and ISPs can jointly provide electrical power and Internet service to users over the existing electrical power distribution network, as suggested in Figure 1.7 [HHL]. Signals may also be carried in the higher-voltage distribution and core transmission facilities of the power grid, but ubiquitous optical communication networks may be more likely to have this responsibility.

The electrical power supply system consists of three network levels: high-voltage network (110 kV or above), medium-voltage network (10–30 kV), and low-voltage network (110/230/440 V). The low-voltage network covers the last few hundreds of meters between the users and the transformer, directly supplying the users served by the last transformer. BoPL employs the low-voltage network as a medium for broadband access. The necessary elements include the PLC base/master station (PLCBS) that couples the Internet with the power supply network and the PLC modem that couples the user with the power line at the customer premise.

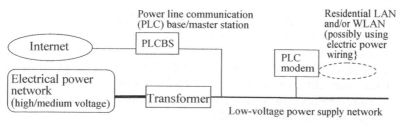

Figure 1.7 Broadband over power line (BoPL) network.

As shown in Figure 1.7, the PLCBS converts the communications signal from the Internet backbone into a format (typically OFDM modulation) that is suitable for transmission through the low-voltage power supply network. The PLC modem converts the communications signal into a standard format for in-residence distribution and provides standard user-side interfaces, such as Ethernet and USB, for different communications devices. The typical data transmission speed of deployed BoPL networks ranges from 256 Kbps to 3 Mbps.

Several standards organizations are developing specifications. The IEEE working groups IEEE P1675 [P1675], IEEE P1775 [P1775], and IEEE P1901 [P1901] are pursuing, respectively, standards on BoPL hardware installation and safety, BoPL EMC and consensus test, and BoPL MAC and physical layer specifications. IEEE P1901 specifies two alternative and incompatible physical layers (PHY), one using FFT-based OFDM and the other wavelet-based OFDM. The HomePlug Powerline Alliance (http://www.homeplug.org), [HOMEPLUG] founded in 2000 by several technology companies for the specifications of BoPL products and services, contributed to P1901.

The European Telecommunications Standards Institute (ETSI) advances BoPL standards through the Power Line Telecommunications (PLT) project [PLT]. This work contributed to the pending ITU-T G.hn standard, which covers home networking on several different transmission media (including power wiring) and thus has only a partial overlap with P1901 that also addresses the access network. The first, the physical layer (PHY) part of G.hn was approved in October 2009.

Concern about the interference caused by BoPL signals radiating from power lines has so far limited the deployment of this technology. Power lines are typically untwisted and unshielded, and are effectively large radiating antennas. Depending on the allocated spectrum, interference with other radio services can be a problem. Conversely, because of the lack of shielding, BoPL signals are also subject to interference from outside radio services. The incompatible PHY alternatives within IEEE P1901 and between P1901 and ITU-T G.hn could also slow deployment. However, the benefits of PLC, including support of the massive Smart Grid (electrical distribution) projects in many nations, may accelerate solutions to these problems.

1.3.4 Broadband Wireless Access (BWA)

Aiming at providing high-speed data access, both direct from fixed or mobile wireless devices and as backhaul from traffic aggregation points, BWA uses licensed and unlicensed spectra over a relatively wide area. The particular BWA technology standardized by the IEEE 802.16 working group is known as WiMAX. According to the IEEE 802.16-2004 standard, broadband means "having instantaneous bandwidth greater than around 1 MHz and supporting data rates greater than about 1.5 Mb/s."

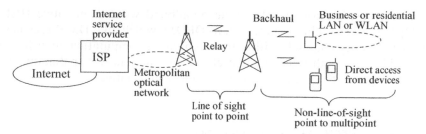

Figure 1.8   IEEE 802.16 WiMAX (broadband wireless access).

The original standard IEEE 802.16a [802.16a] specifies BWA in the 10- to 66-GHz range, which requires direct line-of-sight (LOS) connection in favorable circumstances. IEEE 802.16-2004 [802.16-2004] added support for 2- to 11-GHz non line-of-sight (NLOS) connection between users and the wireless BS, and allows fixed wireless access of up to 70-Mbps data rate and at up to 30-mi service distance.

IEEE 802.16e [802.16e] provides an improvement on the modulation schemes. It enables fixed as well as mobile wireless applications primarily by enhancing its modulation and multiple access system, orthogonal frequency division multiple access (OFDMA). adaptive antenna system (AAS) and multiple input multiple output (MIMO) technology are adopted to improve BWA performance.

LOS transmission uses a fixed dish antenna, on a rooftop or pole, aimed at the WiMAX tower. These directional antennas increase the link gain and support spatial multiplexing. The system uses higher frequencies in the range between 10 and 66 GHz, where there is less interference and higher bandwidth. As shown in Figure 1.8, LOS is suitable for P2P backhaul transmission between WiMAX transmitters as well as from visible access locations. When reaching the customer premises, NLOS transmission is preferred for its capability of better diffraction around obstacles.

Although WiMAX (and the similar Korean Wireless Broadband or WiBro) is increasing in popularity, and there are interesting opportunities for joining WiMAX with PON in a powerful access architecture [YOGC] as described further in Chapter 4, there are alternatives including Europe's High Performance Radio Metropolitan Area Network (HIPERMAN). Other competitors include third-generation (3G) and fourth-generation (4G) cellular mobile systems, most notably 3G long term evolution (LTE) and the coming 4G long term evolution-advanced (LTE-A), but also including Universal Mobile Telecommunications System (UMTS), 1x Evolution-Data Optimized (EV-DO), mesh networked Wi-Fi, and the still developing mobile BWA [IEEE 802.20]. Wireless regional networking exploiting unused capacity in the television broadcast bands, an example of "cognitive radio" that is intended primarily for rural areas, is another interesting broadband wireless initiative [IEEE 802.22]. Some key technologies, such as OFDM, which, like DMT, uses many narrow frequency channels with signals generated computationally using the FFT, and MIMO, with multiple in–multiple out antennas for spatial diversity or capacity gain, are common to many of these alternatives.

1.4 FIBER-OPTIC ACCESS SYSTEMS

PON is one of several alternatives for fiber-optic access networking, and optical access networking is itself only a part of a larger optical networking infrastructure. Component integration and new packaging technologies in optical communications, among other rapid advances in optical technologies, have made fiber-optic communication a promising solution for broadband services at a reasonable cost.

For example, in the core network, commercially available single-mode optical fiber supports transmission at 10 Gbps at distances of over 60 km without repeaters. WDM with 40 Gbps per wavelength transmission was introduced in recent years. Already in 2001 NEC demonstrated an experimental ultra-dense wavelength-division multiplexed (DWDM) system with 40 Gbps per wavelength and a total capacity of 10.92 Tbps [FKMOISOO]. Advances such as these spur the massive deployment of long-haul and metro fiber-optic networks. The development of WDM, using tens of wavelengths over a single fiber, allows the transmission of a huge volume of data over the long-haul network. Metropolitan area networks (MANs) also rely on optical fibers to transmit multiple wavelengths over a relatively shorter range. Figure 1.9 exemplifies a typical network structure where optical fibers are employed to support both long-haul and metro transmission.

Extending optical communications to the access domain is a part of this network evolution that is necessary to deliver new services and applications. There are diverse optical access solutions, depending on how close the fiber comes to the end user. The generic architecture is called FTTx, which includes

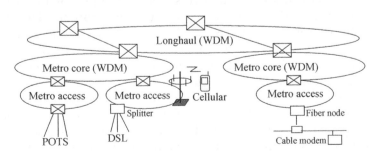

Figure 1.9 Hierarchy of fiber-optic networks.

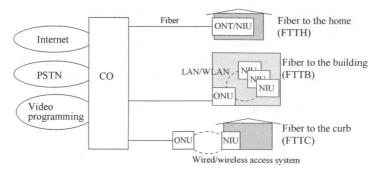

Figure 1.10   FTTx fiber access alternatives.

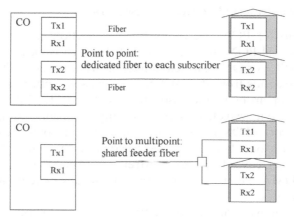

Figure 1.11   Point-to-point and point-to-multipoint fiber access.

FTTH, FTTB, FTTC, FTTCab, and so on [KEISER]. As illustrated in Figure 1.10, FTTH brings an optical fiber-to-the-user's home, while in the FTTB, FTTC, and FTTCab architectures, the optical fiber reaches, respectively, the user's building, neighborhood, or a small cabinet located near the subscribers. In the last three cases, there is an additional distribution network from the ONU to the NIUs, and the signal is converted from optical to electrical to feed the users over copper telephone wires or coaxial cables. FTTC and FTTCab are remarkably similar, in basic architectural concept of fiber partway and copper the rest of the way, to VDSL and HFC.

For those systems offering optical communication all the way to the user premises, P2P and point-to-multipoint (P2MP), shown in Figure 1.11, are two ways of service distribution over the fiber-optic access network. In the P2P connection, each subscriber is connected to a CO through a dedicated fiber. Network upgrade for higher capacity is straightforward and power budget is sufficient for very long link reach. In order to provide the P2P connection, each subscriber requires a separate fiber port in the CO, and the transceiver count has to be two times the number of subscribers.

Unfortunately, for most residential customers, the overall cost of running and managing active components at both ends of a fiber is prohibitive. Without the benefits of large-scale cost sharing as in the backbone network, the access network must strive to minimize cost. Therefore, the alternative of P2MP connection is the favored option. P2MP shares facilities, in particular, one transceiver at the CO and one long feeder fiber, among a group of subscribers,. It simplifies the access infrastructure and is more cost-effective in terms of deployment and maintenance. On the other hand, proper management is required to allocate the shared network resources among the associated subscribers.

1.4.1 PON as a Preferred Optical Access Network

The motivations for PONs were described in Section 1.1. As illustrated in Figure 1.2, a typical PON consists of one OLT, which is located at a CO, and *n* associated ONUs or ONTs, which deliver network traffic to the subscribers. A single fiber extends from the OLT to a 1:*n* passive optical splitter, fanning out *n* single fiber drops to the associated ONUs/ONTs [LA]. Downstream transmission on one wavelength typically time division multiplexes traffic for different users (TDM) and is broadcast to all connected ONUs and ONTs, which pick out their own data. Encryption or secure ONUs may be employed to deter eavesdropping. Upstream transmission on another wavelength uses TDMA. Figure 1.12 illustrates these downstream and upstream transmission modes.

In comparison with alternative optical access networks, PONs, taking advantage of the P2MP architecture and passive optical elements, have a

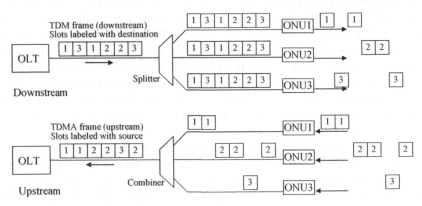

Figure 1.12 Data transmission over TDM-PON. Top: downstream TDM. Bottom: upstream TDMA.

reduced feeder fiber count, fewer transceivers at the CO, no intermediate powering, and aggregate data rates of 2.488 Gbps and more. PON can be feasible where high fiber installation and maintenance costs preclude dedicated P2P connections. It minimizes cost while supporting fine service granularities and scalability. Service coverage is typically for distances up to 20 km.

In addition to potential energy savings from no intermediate powering, additional and substantial energy savings at the ONUs are possible through implementation of a sleep mode [WONG]. ITU-T Recommendation G.sup45 specifies two alternative energy saving modes for GPON ONUs, one controlling only the transmitter and the other, a cyclic sleep mode, controlling both transmitter and receiver. There can be substantial energy savings, but it is important that ONUs can be awakened quickly in order to avoid service disruption. Section 3.4.3 introduces a sleep-control MAC for 10G EPON.

The passive elements of a PON include fiber-optic cables and a passive optical splitter. The splitter allows the downstream traffic from the OLT and the upstream traffic to the OLT to be split from and combined onto the shared portion of the fiber. Less expensive and longer-lived passive components are employed in PON, replacing the active electronic components such as regenerators, repeaters, and amplifiers in DSL or cable data systems.

The OLT and ONUs/ONTs are the active elements, located at the easily serviceable endpoints of a PON. The OLT supports management functions at the CO and is capable of managing tens of downstream links in a PON. ONUs/ONTs are the customer premise equipment (CPE) and have to support only their own link to the CO. As a result, a single ONU/ONT device is relatively inexpensive, while the OLT device tends to be more capable and thus more expensive. Eliminating the intermediate powering makes PONs easier to maintain and lowers overall system cost.

In a typical TDM-PON, different wavelengths are employed for the two directions to avoid collisions and interactions between the downstream and upstream traffic flows. As illustrated in the top (downstream) drawing of Figure 1.12, data are broadcast from the OLT to each ONU/ONT using the entire bandwidth of the downstream channel, and all the downstream data are carried in one wavelength (e.g., 1490 nm). ONUs/ONTs selectively receive frames destined to them by matching the addresses in the received data. Security, as suggested earlier, can be realized through encryption or physically securing the ONUs. The "broadcast and select" architecture supports downstream multimedia services such as video broadcasting.

In the upstream direction, as illustrated in the bottom sketch of Figure 1.12, multiple ONUs/ONTs share the common upstream channel, and another wavelength (e.g., 1310 nm) is employed. Only a single ONU/ONT may transmit during a time slot in order to avoid data collisions. Because of the directional nature of the passive optical splitter, each ONU/ONT transmits directly to the OLT but not to other ONUs. An ONU/ONT buffers the data from the end users until its time slot arrives. The buffered data are transmitted in a burst to the OLT in the exclusively assigned time slot at the full channel speed. The result is that the P2MP architecture of a TDM-PON effectively supports multiple P2P links between the OLT and the associated ONUs/ONTs.

1.5 PON DEPLOYMENT AND EVOLUTION

PON deployment depends on satisfying demand for enhanced broadband access at an acceptable cost of optical access networking. FTTx is favored by most communications operators as the long-term solution for ever-growing bandwidth demands, and among the FTTx alternatives, PON is preferred for its cost/performance advantages. New greenfield PON installations are already an economical way to meet subscriber requirements for broadband services. "Brownfield" migration from existing DSL customers to PON is less optimistic but still attractive enough that major telecommunications operators, particularly Verizon in the United States, are pursuing it vigorously. An evolutionary deployment strategy, one neighborhood at a time, will include program services in addition to Internet access, generating the revenue needed to wire the next neighborhood.

PON technologies are evolving along with optical devices and high-speed networking. Early work on PONs was conducted in the 1980s as a new approach to the last mile problem. BT built a PON demonstration system, the telephony over passive optical network (TPON), to provide both telephony and low-rate data services to multiple subscribers [WV]. In the 1990s, the FSAN group (http://www.fsanweb.org) was formed by service providers to facilitate the creation of suitable access network equipment standards. FSAN announced the specification of BPON in 1998, heralding the first widespread use of FTTx technology. The ITU-T standardization sector soon turned BPON into ITU-T Recommendation G.983.x. Based on ATM protocol, BPON is the first standardized PON technology.

Since then, several different PON standards, including EPON and GPON, have been approved to facilitate broadband access. EPON was introduced by IEEE to explore Ethernet as the encapsulation layer. It initially supported transmission speeds of 1 Gbps and was favored by NTT, which began deploying EPON for broadband access in Japan in 2003. The GPON recommendations were ratified by ITU-T as an extension of BPON. North American carriers such as AT&T and Verizon adopted GPON as the preferred PON technology to roll out their FTTx efforts. The most significant difference between each "flavor" of PON is the supported line rates and the type of bearer packets. Table 1.4 gives an overview of the current PON standards. Chapter 3 will discuss their characteristics in detail.

More recently, work on next-generation passive optical network (XG-PON) has sought to define a PON architecture that is compatible with the current GPON and EPON [EMPF]. It has minimal overlap with the upstream spectrum plans of GPON and EPON, and its downstream wavelength is compatible with video overlay and matches the downstream wavelength in

TABLE 1.4 PON Standards

	BPON	EPON	GPON	10G EPON	XG-PON
Standard	ITU-T Rec. G.983.x	IEEE 802.3ah	ITU-T Rec. G.984.x	IEEE 802.3av	ITU-T Rec. G.987.x
Downstream wavelength	1490 nm	1490 nm	1490 nm	1577 nm	1577 nm
Upstream wavelength	1310 nm	1310 nm	1310 nm	1270/1310 nm	1270 nm
Typical rate	622.08 Mbps DS, 155.52 Mbps US	1 Gbps symmetric	2.488 Gbps DS, 1.244 Gbps US	10 Gbps DS, 1 or 10 Gbps US	9.952 Gbps DS, 2.488 Gbps US

the [IEEE P802.3av] (10GB EPON) draft standard. It may also incorporate dense wavelength division multiplexing (DWDM), allowing the creation of multiple "logical" PONs each utilizing one downstream and one upstream wavelength. The framing and time TDMA control for XG-PON will extend that of GPON, with the hope of convergence with framing and multiple access structures of 10G-EPON. The XG-PON initiative is described in more detail in Chapter 4.

An interesting question of PON evolution is its relationship with other access systems and the public network as a whole. We believe that PON will form a symbiotic relationship with BWA, sharing both access and backhaul functions with 4G cellular mobile and IEEE 802.16 WiMAX in a way that increases the reliability of both wireless and optical segments. For example, PON may backhaul from a WiMAX access point, and WiMAX may provide a protection overlay to PON customers. PON will also provide much of the optical networking infrastructure supporting communication among BSs for coordinated multipoint (CoMP) as part of the 4G LTE-A standard [FUTON]. We will return to these ideas later in this book.

REFERENCES

[AYDA] C. Assi, Y. Ye, S. Dixit, & M. Ali, "Dynamic bandwidth allocation for quality-of-service over Ethernet PONs," IEEE J. Sel. Area. Comm., 21(9), pp. 1467–1477, November, 2003.

[AZZAM] A. Azzam, *High-Speed Cable Modems*, McGraw-Hill, 1997.

[BPCSYKKM] A. Banerjee, Y. Park, F. Clarke, H. Song, S. Yang, G. Kramer, K. Kim, B. Mukherjee, "Wavelength-division-multiplexed passive optical network (WDM-PON) technologies for broadband access: A review [Invited]," J. Opt. Netw., 4(11), November, 2005, available at http://networks.cs.ucdavis.edu/publications/2005_amitabha_2005-11-19_05_24_32.pdf.

[BREUER] D. Breuer, F. Geilhardt, R. Hulsermann, M. Kind, C. Lange, T. Monath, & E. Weis, "Opportunities for next-generation optical access," IEEE Communications Magazine, February, 2011.

[CISCO] "Cisco cloud computing—Data center strategy, architecture, and solutions," White paper for US, public sector, first edition, 2009, available at http://www.cisco.com/web/strategy/docs/gov/CiscoCloudComputing_WP.pdf

[CJMG] J. Cioffi, S. Jagannathan, M. Mohseni, & G. Ginis, "CuPON—The copper alternative to PON," IEEE Communications Magazine, June, 2007.

[DR] A. Dutta-Roy, "An overview of cable modem technology and market perspectives," IEEEE Communications Magazine, June, 2001.

[ECHKLOP] F. Effenberger, D. Cleary, O. Haran, G. Kramer, R.-D. Li, M. Oron, & T. Pfeiffer, "An introduction to PON technologies," IEEE Communications Magazine, March, 2007.

[EMPF] F. Effenberger, H. Mukai, S. Park, & T. Pfeiffer, "Next-generation PON—Part II: Candidate systems for next-generation PON," IEEE Communications Magazine, November, 2009.

[FJ] D. Fellows & D. Jones, "DOCSIS cable modem technology," IEEE Communications Magazine, March, 2001.

[FKMOISOO] K. Fukuchi, T. Kasamatsu, M. Morie, R. Ohhira, T. Ito, K. Sekiya, D. Ogasahara, & T. Ono, "10.92-Tb/s (273/spl times/40-Gb/s) triple-band/ultra-dense WDM optical-repeatered transmission experiment," Proc. Optical Fiber Conference (OFC) 2001, March, 2001.

[FUTON] P. Monteiro, S. Pato, E. Lopez, D. Wake, N. Gomes, & A. Gameiro, "Fiber optic networks for distributed radio architectures: FUTON concept and operation," Proc. IEEE WCNC 2010 Optical-Wireless Workshop, April, 2010.

[GC] G. Ginis & J. Cioffi, "Vectored transmission for digital subscriber line systems," IEEE J. Sel. Area. Comm., 20(5), pp. 1085–1104, 2002.

[GHW] R. Gitlin, J. Hayes, & S. Weinstein, *Data Communication Principles*, Plenum, 1992.

[GREEN] P.E. Green, Jr, *Fiber to the Home: The New Empowerment*, Wiley, 2006.

[HHL] H. Hrasnica, A. Haidine, & R. Lehnert, *Broad band Powerline Communications: Network Design*, Wiley, 2004.

[HOMEPLUG] "HomePlug AV White Paper, HomePlug Powerline Alliance Inc.," 2005, available at https://www.homeplug.org/products/whitepapers/HPAV-White-Paper_050818.pdf.

[IEEE 802.3ah] IEEE Std 802.3ah-2004, "Ethernet in the First Mile," available at http://www.ieee802.org/3/purchase/index.html.

[IEEE 802.3av] IEEE draft Std 802.3av-2009, "10Gb/s Ethernet Passive Optical Networks," available at http://standards.ieee.org/getieee802/download/802.3av-2009 .pdf.

[IEEE 802.20] M. Klerer, "Introduction to IEEE 802.20," available at http://www.ieee802.org/20/P_Docs/IEEE%20802.20%20PD-04.pdf.

[IEEE 802.22] "Enabling rural broadband wireless access using cognitive radio technology in TV whitespaces," IEEE 802.22 Project Authorization Request (modified), December 9, 2009, available at http://www.ieee802.org/22.

[ITU-T G.983.x] "Broadband optical access systems based on passive optical networks (PON), International Telecommunications Union 2005," available at www.itu.int/rec/T_REC-G.983..x/en, where the number (1,2,3,...) of a component standard should replace "x."

[ITU-T G.984.x] "Gigabit-capable passive optical networks (GPON): General characteristics, International Telecommunications Union 2003," available at www.itu.int/rec/T_REC-G.983..x/en, where the number (1,2,3,...) of a component standard should replace "x."

[KEISER] G. Keiser, *FTTX Concepts and Applications*, Wiley-IEEE Press, 2006.

[LA] Y. Luo and N. Ansari, "LSTP for dynamic bandwidth allocation and QoS provisioning over EPONs," OSA J. Opt. Netw., 4(9), pp. 561–572, 2005.

[LAM] C. Lam (Ed.), *Passive Optical Networks—Principles and Practice*, Academic Press, 2007.

[MAIER] M. Maier, "WDM passive optical networks and beyond: The road ahead," IEEE J. Opt. Commun. Netw., 1(4), September, 2009.

[ODMP] F. Ouyang, P. Duvaut, O. Moreno, & L. Pierrugues, "The first step of long-reach ADSL: Smart DSL technology, READSL," IEEE Communications Magazine, September, 2003.

[OFDMPON] N. Cvijetic, D. Qian, T. Wang, & S. Weinstein, "OFDM for next-generation optical access networks," Proc. IEEE WCNC 2010 Optical-Wireless Workshop, April 2010.

[PLT] "Powerline telecommunications (PLT): Coexistence of access and in-house powerline systems," ETSI ES 201 867 v1.1.1, November, 2000, draft available at http://www.etsi.org/deliver/etsi_es/201800_201899/201867/01.01.01_50/es_201867v010101m.pdf.

[SHUMATE] P. Shumate, "Fiber-to-the-home: 1977–2007," J. Lightwave Technol., 26(9), May, 2008.

[SSCP] T. Starr, M. Sorbara, J. Cioffi, & P. Silverman, *DSL Advances*, Prentice-Hall, 2002.

[WEIN] S. Weinstein, *The Multimedia Internet*, Springer, 2005.

[WONG] S. Wong, L. Valcarenghi, S.-H. Yen, D. Campelo, S. Yamashita, & L. Kazovsky, "Sleep mode for energy saving PONs: Advantages and drawbacks," 2nd International Workshop on Green Communications, IEEE Globecom 2009, Honolulu, Dec. 4, 2009.

[WV] J. Walrand & P. Varaiya, *High-Performance Communication Networks*, Morgan-Kaufmann, 2000.

[YOGC] K. Yang, S. Ou, K. Guild, & H.-H. Chen, "Convergence of Ethernet PON and IEEE 802.16 broadband access networks and its QoS-aware dynamic bandwidth allocation scheme," IEEE JSAC, 27(2), February, 2009.

2

PON ARCHITECTURE
AND COMPONENTS

2.1 ARCHITECTURAL CONCEPTS AND ALTERNATIVES

Passive optical networking is a full duplex technology that uses inexpensive optical splitters (in the downstream direction) to divide a single fiber coming from the backbone network into separate drops feeding individual subscribers in the access network. The standardized passive optical networks (PONs) employ point to multipoint (P2MP) as the basic communication architecture, realized in the splitter in Figure 1.2, where the optical line terminal (OLT) is the control point for the entire PON and the optical network units (ONUs)/ optical network terminations (ONTs) are the centrally controlled end (client) points. All downstream traffic is broadcast to all of the end nodes, each of which admits only that traffic destined for itself. The time division multiple access (TDMA) reverse (upstream) traffic is effectively point to point (P2P), with data from one endpoint transmitted to one OLT.

2.1.1 Topologies

Although PONs may exhibit diverse network topologies as discussed below, the P2MP physical system supports a logical tree architecture, in which an OLT is passively linked to the associated ONUs/ONTs through a passive optical

The ComSoc Guide to Passive Optical Networks: Enhancing the Last Mile Access,
First Edition. Stephen Weinstein, Yuanqiu Luo, Ting Wang.
© 2012 Institute of Electrical and Electronics Engineers. Published 2012 by John Wiley & Sons, Inc.

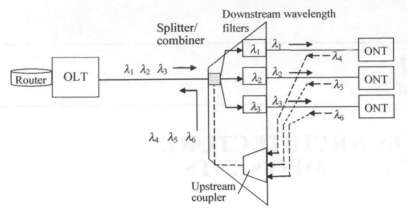

Figure 2.1   Passive unicasting using distinct wavelengths for different ONTs.

splitter. Downstream optical signals are split into multiple fibers at the splitter, and upstream optical signals are combined, usually through TDMA, onto a single upstream fiber. Section 1.4.1 already introduced the advantages of PON in comparison to legacy P2P networks.

Although P2MP in its wiring topology, the downstream system can become a *logical* P2P system, for example, by the use of multiple wavelengths. Passive wavelength filters in the splitter could restrict multicasting of each wavelength to a small group of end users or even restrict one wavelength to one user as illustrated in Figure 2.1. This figure also suggests the possibility of dedicated wavelengths in the upstream direction, as an alternative to TDMA on a single wavelength. Although separate wavelengths are shown for downstream and upstream transmissions associated with a given ONT, the upstream wavelength can be the same as the downstream wavelength, using the downstream signal to generate the upstream carrier signal in the interest of saving wavelengths, as described in Chapter 4. Different modulation types can separate the signals. If finer-grain channels are needed, subcarrier modulation is possible. A frequency division multiplexed signal, that is, a group of modulated subcarriers, is itself modulated onto a particular wavelength. This conserves wavelengths and supports a large number of P2P connections. This is also discussed in Chapter 4.

Although a tree topology is presumed in most of this book, Figure 2.2 illustrates an actual choice among three fundamental topologies: tree, bus, and ring. In the tree topology, the associated (say, N) ONUs/ONTs are located in a relatively small range from the OLT. The distances from the (single) splitter to the N ONUs/ONTs are similar, and one 1:N splitter evenly distributes the signals. The tree topology suits an urban area, where the subscribers are closely located.

The bus topology considerably extends the spacing of a group of ONUs/ONTs served by a particular OLT. It suits relatively rural areas, where population density is low and subscribers are located far from each other.

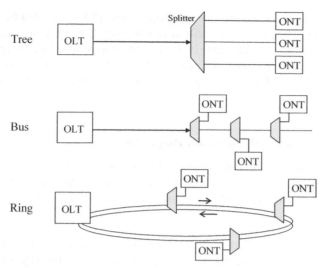

Figure 2.2 Basic topologies. Fibers from OLT carry both downstream (arrow) and upstream traffic. The ring typically employs two fibers, operating in opposite downstream directions, one for normal use and one for protection [YEH].

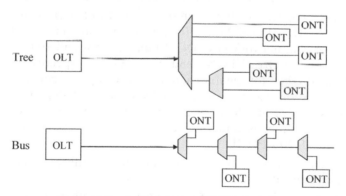

Figure 2.3 Adding a new ONU to a PON.

The ring topology is essentially an extension from the bus topology with two trunk fibers. It provides failure protection to the PON service as each trunk fiber can back up the other when fiber or certain component failures occur. An example of a hybrid wavelength division multiplexing (WDM) ring-tree PON is described in Chapter 4.

All three topologies are extensible to more than the original set of users, as suggested in Figure 2.3. The tree topology can cascade a second layer of PON tree with a second splitter. The bus and ring topologies simply attach a new splitter to the trunk or ring fiber.

Although basically a local access mechanism, PON can also be used over large distances. This, too, is described in Chapter 4, as part of a view toward PON even beyond the current 10-Gbps developments. The technical community and manufacturers aspire to a system with a 100-km reach, a per-user data rate of at least 1 Gbps, and up to 1000 users, together defining a huge capacity [ELBERS].

2.1.2   Downstream and Upstream Requirements

The precise downstream and upstream requirements for PONs depend on the particular choice among the standard types defined in the next subsection and how they are configured. These standard types all share requirements for the following:

- A single wavelength, usually 1490 nm, for data transmission in the downstream direction and another single wavelength, usually 1310 nm, for upstream transmission. A second downstream wavelength, usually 1550 nm, may be provided for downstream video.
- A node, between the OLT in the serving office and the ONT on customer premises, that contains a passive splitter for downstream transmission and a passive N:1 coupling for the TDMA upstream traffic.
- Aggregate capacities, shared among a group of users (e.g., 32 users), of 622.08 Mbps/155.52 Mbps (downstream/upstream) for broadband passive optical network (BPON), 1 Gbps/1 Gbps and 10 Gbps/10 Gbps for Ethernet PON (EPON) and 10G-EPON, respectively, and 2.488 Gbps/1.244 Gbps for gigabit-capable passive optical network (GPON). next-generation passive optical network (XG-PON), with 10 Gbps/2.5 Gbps service in a first generation and 10 Gbps/10 Gbps in a subsequent generation, is under development as an extension of GPON [ECHKLOP, XG-PON].
- A medium access control (MAC) facilitating efficient sharing of the upstream capacity.
- Means for ensuring quality of service (QoS) for different classes of traffic and for configuration, fault, and performance management of system elements.

These requirements are further explained later in this chapter and in Chapter 3, where the detailed functions of BPON, GPON, and EPON are described.

2.1.3   BPON, GPON, and EPON Systems

We offer here introductions to these three standard PON types [EFFEN] and their 10G extensions, with more detailed discussion of standards in the next chapter. Table 2.1 compares them in terms of data rates, optical distribution

TABLE 2.1 Overview of BPON, GPON, XG-PON, EPON, and 10G-EPON

	BPON	GPON	XG-PON	EPON	10G-EPON
Classes	ITU-T G.983 B, C	ITU-T G.984 A, B, C	ITU-T G.987 N1, N2, E1, E2	IEEE 802.3ah PX10, PX20	IEEE 802.3av P(R)X10 P(R)X20 P(R)X30
Downstream	155.52 Mbps 622.08 Mbps	1.244 Gbps 2.488 Gbps	9.952 Gbps	1 Gbps	1 Gbps 10 Gbps
Upstream	155.52 Mbps 622.08 Mbps	155.52 Mbps 622.08 Mbps 1.244 Gbps 2.488 Gbps	2.488 Gbps	1 Gbps	1 Gbps 10 Gbps

Min/Max Loss of ODN Classes

ODN Class	Min/Max Loss (dB)		ODN Class	Min/Max Loss (dB)
A	5/20		E2	20/35
B	10/25		PX10	5/20
C	15/30		PX20	10/24
N1	14/29		P(R)X10	5/20
N2	16/31		P(R)X20	10/24
E1	18/33		P(R)X30	15/29

From http://www.answers.com/topic/pon, http://pdfserv.maxim-ic.com/en/an/AN1102.pdf, and other sources.

network (ODN) classes, and the transmission loss bounds for those classes. The maximum number of ONTs for each class depends on many factors, including transmission loss, fiber distance, and splitter loss, and parameters such as optical power budget, optical link penalty, Tx launch power, and Rx sensitivity, and so is not specified here.

As noted earlier, BPON began in 1995 in the Full Service Access Network (FSAN) industry consortium, initially consisting of BellSouth, British Telecom, Deutsche Telekom, France Telecom, and Nippon Telegraph and Telephone (NTT) Company [EFFEN]. It began as asynchronous transfer mode passive optical network (APON) and still is often viewed as an asynchronous transfer mode (ATM)-PON, based on encapsulation of all types of traffic into ATM cells. ATM, a switched connection-oriented service, offers advantages of switching efficiency and assured QoS but has drawbacks in its relatively high overhead (ratio of the 5-byte header to the 53-byte total cell size), need to convert to Ethernet at most user locations, and awkwardness in accommodating Internet protocol (IP) traffic.

The G.983 standard for BPON that FSAN sponsored began, as G.983.1 in late 1998, with 155/155 (megabits per second downstream and upstream) and

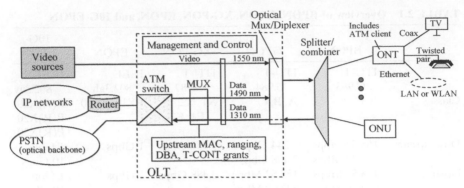

Figure 2.4 ATM-based BPON supporting triple play services. T-CONT, transmission containers.

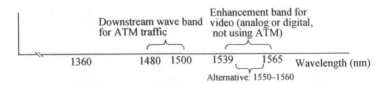

Figure 2.5 Revised waveband allocation in G.983.3 to support a downstream video channel [IDA-BB].

622/155 systems and evolved to higher speeds; enhanced security features; management and control functions for voice, data, and video; inclusion of additional wavelengths particularly for video broadcasting; flexible upstream bandwidth provisioning such as dynamic bandwidth allocation (DBA); and other features enumerated in the standards overview in Chapter 3. Figure 2.4 illustrates the major functions of an ATM-based BPON, and Figure 2.5 the change in frequency plan, made in G.983.3, to support the downstream video services channel shown in Figure 2.4. This channel can carry analog and/or digital video, directly passed and not mapped into an ATM stream.

Note the important ranging function, found in all the standard PONs and in many other access systems, to assure that TDMA bursts from different users arrive at their assigned times. Ranging is the process of determining the propagation (and processing) delay characterizing the channel between the OLT and a particular ONU/ONT.

GPON originated in FSAN in 2001. The standardization process produced, in 2003–2001, International Telecommunication Union-Telecommunication (ITU-T) Standard series G.984.x covering the basic architecture, the physical medium dependent layer, the transmission convergence layer, and management requirements [ECHKLOP]. The motivations for GPON were needs for

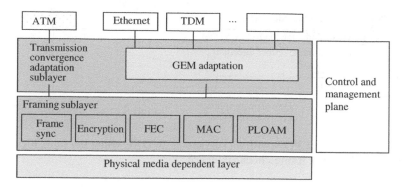

FEC: forward error correction
MAC: medium access control
PLOAM: physical layer operation, administration, and maintenance

Figure 2.6 GPON protocol layers.

larger capacity, greater reach, higher split ratio, and more flexibility in transmission formats. In addition to ATM, other inputs including Ethernet and TDM can be accepted. The typical transmission rates are 2.488 Gbps downstream and 1.244 Gbps upstream.

Figure 2.6 illustrates the protocol layers and particularly the critical role of the GPON encapsulation method (GEM). GEM, described further in the next chapter, adapts the various transmission formats into the GPON transmission frame with a minimum of overhead, providing an efficient means for local distribution of these alternative formats. The GPON transmission convergence (GTC) layer in which GEM is implemented also furnishes MAC and the timing arrangements for ONU transmissions so that they do not interfere in the TDMA upstream channel.

The control and management plane, summarized in Section 2.3 and described in more detail in Chapter 3, includes functions such as registering and monitoring ONUs and configuring bandwidth, data encryption, and forward error correction. Bandwidth allocation can be either static or dynamic, with the latter implemented as either "status-reporting DBA" based on information fed back by ONUs or "non-status-reporting DBA," in which OLTs monitor upstream channel utilization. Management of the GPON, specified in G.984.4, utilizes an ONU management information base (MIB) and the ONT management control channel protocol. BPON and GPON use a standard ONT management and control interface (OMCI) between the OLT and ONU/ONT, thus enabling networks with components from different vendors.

EPON was developed by the Institute of Electrical and Electronics Engineers (IEEE) 802.3 "Ethernet in the First Mile" Study Group, formed in late 2000, with the mandate to extend Ethernet, long prevalent in wired LANs, into the subscriber access network. It takes advantage of the same

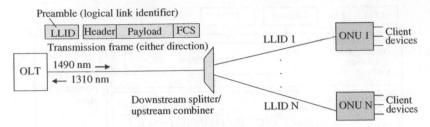

Figure 2.7 Simplified EPON with all services at an ONU using the same logical link identifier (LLID). FCS, frame check sequence.

frame format and MAC as that of standard Ethernet switches or any Ethernet component. The study group's work resulted in the approval of IEEE 802.3ah in 2004.

IEEE 802.3ah addresses the physical and MAC (data link) layers in which Ethernet transmission is defined. Transmission at 1 Gbps in each direction occurs on a 1490-nm optical carrier downstream and on a 1310-nm carrier upstream, with 1550 nm reserved for other uses such as analog video broadcasting. There is no encapsulation framing in either direction. Figure 2.7 shows the essential elements, with a more complete description in Chapter 3.

2.1.4 Medium Access Techniques

Medium access, in this and in most contexts, refers to the admission of end users to a shared upstream channel with either resolution or avoidance of contention. In such a situation, medium access is required for either connection-oriented communication, in which signaling establishes the path, or connectionless communication, where frames or packets are forwarded without prior establishment of a path.

There are two principal modes of medium access: distributed contention and central allocation. In distributed contention access, users compete among themselves for the shared time/bandwidth resources, either by resolving collisions in a distributed fashion, for example, the original "transmit first, then back off a random amount if there is a collision" method of carrier sense multiple access–collision detection (CSMA-CD), or one of the several carrier sense multiple access–collision avoid (CSMA-CA) methods of "listen first before sending, but if a collision occurs anyway, then back off a random amount to avoid another."

Request–allocation schemes, widely used in cable data systems and in cellular mobile and Worldwide Interoperability for Microwave Access (WiMAX) wireless systems as well as PON, offer end users a request channel that the central controller (e.g., the OLT) responds to with time slot and/or frequency band allocations. An allocation system does not necessarily require an end

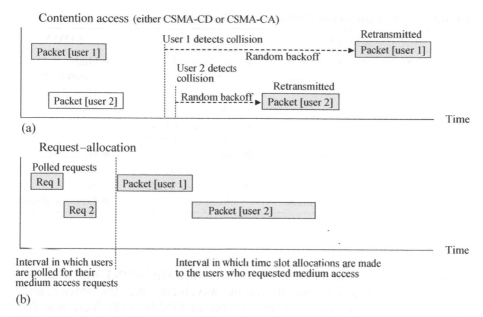

Figure 2.8 Basic medium access modes of contention access and request–allocation.

user request; allocations may be made on the basis of management policy or prior configuration. Figure 2.8 is a simplified illustration of time slot allocations for the distributed contention access and request–allocation protocols. Note that "packet" may be replaced by "frame" if the basic transmission unit is at layer 2 (link) rather than at layer 3 (network).

Different kinds of traffic require different priorities. For example, voice over Internet protocol (VoIP) packets or frames, because of the severe requirement for low delay, may be given transmission priority. On shared access facilities, this may be accomplished in contention access by letting VoIP packets select backoff from a shorter random interval, and in request–allocation access, this may be accomplished by giving earlier and more regular slot allocations. This is one of many approaches to providing class of service (COS) differentiation of different types of traffic. In allocation systems, priority may be given by earlier, more regular, and larger allocations of available time or frequency (wavelength) resources.

As an illustrative example, the EPON multipoint control protocol (MPCP) can provide COS-based medium access, allowing multiple ONUs/ONTs to share the same feeder fiber using prioritized allocations [KM1]. Alternative medium access technologies, that is, forms of physical-level channel sharing, are possible, namely, code division multiple access (CDMA), TDMA, and wavelength division multiple access (WDMA) [KM2]. These are described below and compared in Table 2.2.

TABLE 2.2 Comparison of Medium Access Technologies

	CDMA	TDMA	WDMA
Pros	▶ High data security ▶ No fixed limit on the number of ONUs/ONTs	▶ One transceiver at the OLT ▶ Easy to add a new ONU/ONT ▶ Finer granularity than a wavelength	▶ High bandwidth ▶ High data security
Cons	▶ High interchannel interference ▶ High cost with expensive equipment	▶ Needs synchronization ▶ Needs bandwidth allocation	▶ High cost for transmitter array ▶ Low scalability

CDMA encodes the data associated with each ONU/ONT with a unique wideband multiplicative pseudorandom waveform, and the interchannel interference (ICI) increases as the number of ONUs/ONTs increases. The network components must be able to handle signal rates much higher than the carried data rate and deal with the complexities of ICI, thereby making the network cost relatively high. CDMA-PON can be used for radio over fiber supporting CDMA base stations [KIM]. The high cost of CDMA systems is the main concern for its practical applicability in PON.

TDMA, which divides the total capacity of the shared feeder fiber into time slots, is simpler. ONUs/ONTs are synchronized to transmit in exclusive time slot allocations, as illustrated in Figure 2.8. Only one transmitter is needed at the OLT to multiplex the upstream data no matter how many ONUs/ONTs are connected, and a new ONU/ONT can be easily added/dropped. In order to minimize access network cost and to facilitate access network scalability, TDMA and time division multiplexing (TDM) were consistently chosen by the PON standard bodies (IEEE, 2004; ITU-T, 1998; ITU-T, 2001; ITU-T, 2003a; ITU-T, 2003b; ITU-T, 2004a; ITU-T, 2004b) for upstream and downstream data transmission, respectively. Note that these original standards, explained in more detail in Chapter 3, specify only a single wavelength in each direction. Section 2.3.1 provides an introduction to the control mechanism for TDMA resource allocation.

In WDMA, each ONU/ONT transmits its data to the OLT using a specific allocated wavelength. The OLT has a multiwavelength transmitter array to support multiple ONUs/ONTs. The bandwidth provided to each ONU/ONT is determined by the capacity of the assigned wavelength channel, which can vary depending on the interface speeds and modulation types supported by the system. The anticipated commercial implementation of WDMA in PONs is an enhancement meeting the need for increased access network bandwidth. WDM-PONs are described Chapter 4.

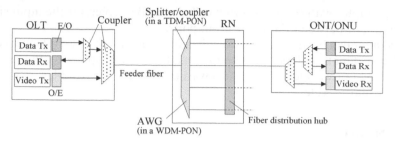

Figure 2.9   Passive splitter/coupler in a TDM-PON and AWG in a WDM-PON.

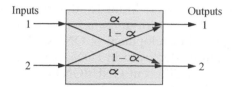

Figure 2.10   A 2 × 2 coupler.

2.2   PASSIVE AND ACTIVE PON COMPONENTS

The various PON systems use standard components, passive (not powered) or active, described in this section. As shown in Figure 2.9, the main passive optical components are the passive optical splitter/coupler (also called a splitter/combiner) and the arrayed waveguide grating (AWG), defined below. In a TDM-PON, the remote node (RN) contains a splitter/couple that is operated in different modes for downstream and upstream traffic. It is a splitter for downstream traffic and a coupler for upstream traffic. In a WDM-PON, the AWG will typically replace the splitter/coupler in the RN for downstream transmission of individual wavelengths or wavelength groups to different end ONT/ONUs, and upstream coupling into one fiber of everything from the ONT/ONUs.

2.2.1   Passive Optical Coupler

A passive optical coupler, part of a passive splitter/coupler, combines multiple signals from different sources into one or more output fibers. Figure 2.10 illustrates a 2 × 2 coupler with two input ports and two output ports. For the signal from input port 1, a fraction α, $0 < \alpha < 1$, is coupled to output port 1, and the remainder is coupled to output port 2. The coupler distributes the signal from input port 2 in the same manner, with the fraction α coupled to output port 2 and the fraction $(1 - \alpha)$ coupled to output port 1. As a result, the signals from both output ports are combinations of the two input signals.

A 2×2 coupler with $\alpha = 0.5$ is often called a 3-dB coupler, for the input signal is split in half between the two output ports.

A coupler can be made either wavelength independent or wavelength selective. A wavelength-independent coupler has the same coupling coefficient over a wide range of wavelengths, while a wavelength-selective coupler only works for particular wavelengths.

2.2.2 Splitter

Inverse to a coupler, a $1 \times N$ splitter is a passive device that splits the optical signal carried by a single input fiber into N output fibers, as suggested in Figure 2.9. The distribution fibers following a PON splitter deliver the same signal from the OLT to the ONUs/ONTs located at different sites. The input optical signals are usually distributed with uniform power over the N output fibers, simplifying the design and development of ONUs/ONTs. Note, as shown in Figure 2.9, that the splitter deployed in PONs is actually a splitter/coupler, combining signals from the different ONUs/ONTs in the upstream direction as well as splitting the downstream signal.

The development of fused fiber technology [AGRAWAL] dramatically reduced passive optical splitter cost and made PONs more attractive. Splitter loss is determined by the number of ports. An optical splitter/coupler introduces almost the same loss to both upstream and downstream transmission. There are multiple ways to realize a specified split ratio, in one splitter or the cascade of several splitters. Figure 2.11 illustrates three ways to achieve a 1:16 split ratio. As the numbers of splitters in a PON increases, the received power decreases proportionally, making 1:64 an effective maximum. IEEE Standard 802.3ah defines split ratios of up to 16. ITU-T Recommendation G.983 specifies split ratios of up to 32, and ITU-T Recommendation G.984 allows split ratios up to 64.

Table 2.3 exemplifies the typical PON commercial splitter/coupler specification. Wavelength range indicates the supporting wavelengths of the splitter and coupler, respectively. Insertion loss (IL) is the amount of signal lost in the total transit through the device and is specified in decibels. The typical IL is

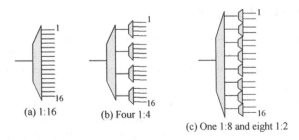

(a) 1:16   (b) Four 1:4

(c) One 1:8 and eight 1:2

Figure 2.11   Alternative 1:16 splitter implementations.

TABLE 2.3 Specification of Optical Passive Splitters

Dual Window Wideband	Wavelength Range (nm)	Typical IL (dB)	Max IL (dB)	Uniformity (dB)	Return Loss (dB)	Directivity (dB)	PDL (dB)
1 × 2 Splitter	1260–1360 and 1480–1580 nm	3.20	3.60	0.70	≥55	≥50	0.20
2 × 4 Splitter	1260–1360 and 1480–1580 nm	6.30	7.80	≥1.6	≥55	≥50	0.30
2 × 8 Splitter	1260–1360 and 1480–1580 nm	10.2	11.00	2.00	≥55	≥55	0.20
1 × 4 Splitter	1260–1360 and 1480–1625 nm	6.60	7.00	0.80	≥55	≥55	0.20
1 × 8 Splitter	1260–1360 and 1480–1625 nm	9.80	10.40	≥1.00	≥55	≥55	0.20
1 × 16 Splitter	1260–1360 and 1480–1625 nm	10.2	11.00	2.00	≥55	≥55	0.20
1 × 32 Splitter	1260–1360 and 1480–1625 nm	15.70	17.10	≥1.50	≥55	≥55	0.30
2 × 16 Splitter	1260–1360 and 1480–1625 nm	13.30	14.50	2.30	≥55	≥55	0.40
2 × 32 Splitter	1260–1360 and 1480–1625 nm	17.00	17.50	2.50	≥55	≥55	0.40

Courtesy of Corning Inc., http://catalog2.corning.com/CorningCableSystems/media/Resource_Documents/product_family_specifications_rl/EVO-411-EN.pdf.

the average IL value measured at the specified center wavelength at room temperature, while Max IL is the maximum IL value for the entire operating wavelength and temperature range. Uniformity is the maximum difference between the leg of the splitter or coupler with the highest IL and the leg with the lowest IL. Polarization-dependent loss (PDL) is the variation of the splitter or coupler's IL with varying states of the signal polarization. Return loss (RL) indicates the amount of power that is reflected back from the input port and thus lost from the forwarded signal. Directivity is defined as the ratio of the power radiated in the output direction to the average of the power radiated in all directions and is measured in dB.

2.2.3   Arrayed Waveguide Grating (AWG)

The AWG, functionally illustrated in Figure 2.12, is capable of multiplexing a group of wavelengths into a single optical fiber as well as separating the signals carried on different wavelengths in a single optical fiber into different output ports. In WDM-PONs, where ONUs/ONTs are served by different wavelengths, AWGs are used as multiplexers/demultiplexers. Section 4.2.2 ("WDM Devices") contains further information about the AWG in addition to the discussion of other components of a WDM-PON.

The operation of an AWG is suggested in Figure 2.13. The incoming signals from a single fiber (labeled 1) traverse an optical splitter (labeled 2) and enter a set of parallel waveguides with different lengths (labeled 3). These waveguides are uncoupled, that is, far enough apart so that the electromagnetic field in one guide does not extend into any other guide. When exiting the waveguide area, the signals traverse an optical combiner (labeled 4) and interfere when entering the output waveguides (labeled 5). The waveguides and free spaces

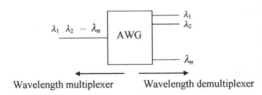

Figure 2.12   Functionality of an arrayed waveguide grating (AWG).

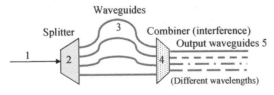

Figure 2.13   AWG operation.

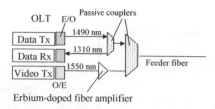

Figure 2.14   The active optical line terminal (OLT).

in AWGs are designed such that each output port receives only one wavelength. An AWG functions as a wavelength demultiplexer by using the lightpath $1 \rightarrow 2 \rightarrow 3 \rightarrow 4 \rightarrow 5$ and as a wavelength multiplexer by using the lightpath $5 \rightarrow 4 \rightarrow 3 \rightarrow 2 \rightarrow 1$.

Despite the label "passive," a PON also has important active components, constituting the OLT and the client ONUs/ONTs.

2.2.4   Optical Line Termination (OLT)

An OLT (Figure 2.14) is the service provider endpoint in a PON, located at the central office (CO) in a typical telco installation. It typically provides digital data and voice transmission using a 1490-nm laser transmitter, analog video transmission using a 1550-nm laser transmitter, and digital data and voice reception in a receiver employing a 1310-nm detector.

The OLT often transmits the downstream data and voice signal through the 1490-nm wavelength using a distributed feedback (DFB) laser. The upstream data and voice signal from the associated ONUs/ONTs are received using a 1310-nm detector. The analog video signal is translated to the 1550-nm wavelength optical signal at the OLT. An erbium-doped fiber amplifier [EDFA] may be employed to amplify the converted video signal before delivering it downstream to the subscribers. Figure 2.15 pictures commercial examples of an OLT and an ONU.

2.2.5   ONU/ONT

An ONU/ONT (Figure 2.16) is the endpoint in a PON, usually located at a subscriber's premises but possibly in the field at, for example, a Wi-Fi or WiMAX access point. The term ONU usually designates field units that may serve more than one network termination, while ONT applies only to a termination point. An ONU/ONT typically provides digital data and voice transmission from a 1310-nm laser transmitter, digital data and voice reception in a 1490 nm detector receiver, and analog video reception in a receiver with a 1550-nm detector.

The upstream data and voice signal of 1310 nm wavelength is transmitted via a directly modulated uncooled Fabry–Perot laser, while the downstream

Figure 2.15 NEC G-EPON OLT (ME2200) and ONU.

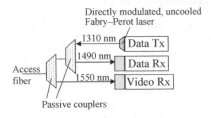

Figure 2.16 ONT/ONU.

TABLE 2.4 Active Equipment Characteristics

Equipment	Service	Laser	Wavelength (nm)
OLT	Data and voice (downstream)	DFB	1480–1500 (typical: 1490)
	Analog video (downstream)	DFB	1550–1560 (typical: 1550)
ONU/ONT	Data and voice (upstream)	Fabry–Perot	1260–1360 (typical: 1310)

data and voice signal is received in a receiver with a 1490-nm detector. The analog video signal is received in a receiver with a 1550 nm detector.

Table 2.4 lists the major characteristics of the OLT and ONU/ONT as specified in the PON standards. In order to carry emerging bidirectional digital services such as video conferences and video blogs, future PONs will need to schedule preferential low-delay upstream transmissions from the ONU/ONT.

2.3 MANAGEMENT AND CONTROL ELEMENTS

Management and control encompasses both normal resources allocation, particularly bandwidth allocation as described in the next section, and the detection of problems or unusual conditions that require attention as briefly reviewed in Section 2.3.4. As is often the choice in communication networks, implementations can be either centralized (e.g., run from a CO) or distributed, using active modules in ONU/ONTs. Centralized control seems preferable from several perspectives, especially for monitoring of the physical channel, where a complete fiber cut makes it impossible to gather information from an ONU [RAD]. Of course, bandwidth allocation requires working communication with a functioning ONU in order to receive capacity requests and upstream data transmissions.

2.3.1 Bandwidth Allocation

In the upstream direction from the ONU/ONT to the OLT, multiple ONUs/ONTs transmit data to the OLT through the shared feeder fiber and the common splitter. Because of the directional nature of the splitter, each ONU/ONT transmits directly to the OLT but not to other ONUs/ONTs. Therefore, the ONUs/ONTs are unable to detect data collision in the upstream direction, and contention-based mechanisms for bandwidth sharing (Section 2.1.4) are difficult to implement in PONs. Moreover, contention-based mechanisms cannot provide guaranteed bandwidth to each ONU/ONT. As a result, in the currently standardized PONs, the OLT arbitrates the upstream capacity allocation among multiple ONUs/ONTs sharing a TDMA channel.

An easy way to share is to assign one or more fixed time slots to each ONU/ONT, which is static bandwidth allocation (SBA). Although SBA is appropriate for constant bit rate (CBR) traffic, it tends to underutilize upstream bandwidth when the traffic is bursty. For example, an "idle" ONU/ONT will occupy the upstream channel for its assigned time slot (in a six-slot frame) even if there are no data to transmit, while the fixed time slots assigned to other ONUs/ONTs may not be long enough to transmit all of their queued-up data. This increases delay for all the data buffered in the "busy" ONUs/ONTs. Many data packets could be backlogged in the buffers of busy ONUs/ONTs, while the upstream channel is lightly loaded or even idle. For this reason, SBA is not preferred as it exacerbates the access network transmission bottleneck.

In order to increase bandwidth efficiency, DBA for upstream traffic in PONs is highly desired. Instead of allocating a fixed time slot to each ONU/ONT, the OLT arbitrates the upstream bandwidth based on the instantaneous demand from each ONU/ONT. DBA results in improved bandwidth utilization. Figure 2.17 shows the concepts of both SBA, with dedicated slot assignments, and DBA, filling the PON upstream time slots as much as possible through constantly changing allocations. In the example of Figure 2.17, DBA

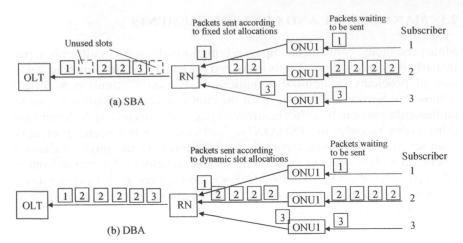

Figure 2.17 Static bandwidth allocation (SBA) and dynamic bandwidth allocation (DBA).

employs the idle time slots from ONU1 and ONU3 to transmit heavy data from ONU2.

Without specifying the DBA algorithm, the PON standards provide the framework and mechanism to facilitate its implementation. Chapter 3 will cover the DBA-related issues defined in the standards.

2.3.2 Quality of Service (QoS)

Beyond upstream bandwidth allocation among different ONUs/ONTs, a major challenge for PONs is to provision QoS appropriate to each traffic class in order to support flourishing new applications. The concept of QoS is long established for voice traffic in the public switched telephone network (PSTN) but relatively new for the IP traffic of the Internet. With the expansion of applications, more QoS-sensitive traffic is carried over the network, including VoIP, online gaming, television distribution in IP packets (Internet protocol television [IPTV]), video on demand (VoD), and multimedia blogs. These applications have different requirements on transmission delay, bandwidth, and data loss, making QoS an urgent concern of network operators and service providers. The metrics to measure QoS quantitatively include delay, bandwidth, packet loss rate, and packet arrival time jitter.

From the broadband access service providers' perspective, QoS-based value-added services are desperately needed to increase their revenue. The primary goal is to improve the overall utility (and hence the revenue) of the PON network by granting priority to higher-value or more performance-sensitive traffic. Implementation of QoS control in a PON does not imply using more bandwidth; it only adds the capability to intelligently allocate the transmission resources.

TABLE 2.5 Bandwidth Types Specified in ITU-T G.983

Bandwidth Types	Priority	Features	Bandwidth Allocation
Fixed	High	Guaranteed and reserved	SBA
Assured	Middle to high	Guaranteed	DBA
Nonassured	Middle	Serve when available	DBA
Best effort	Low	Serve when available, after nonassured	DBA

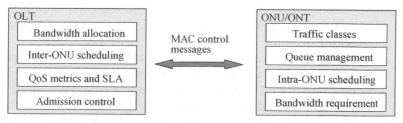

Figure 2.18 QoS provisioning in BPONs.

"Priority" implies either lower drop probability or preferential scheduling when the PON network is congested. The PON standards specify the priority order for bandwidth types, shown in Table 2.5, which are appropriate for different traffic classes. Traffic classification and DBA takes into consideration both the service requirement and the contracted level of service in the service level agreement (SLA). An SLA may, for example, specify peak and average constraints on the amount of traffic in the different bandwidth types that an end user can send or receive.

The usual method to enable QoS is to classify the traffic flows from each ONU/ONT into different classes, buffering them with priority queuing. Traffic classes are mapped into bandwidth types. The ONU/ONT performs traffic policing to control packet drop to avoid buffer overflow. Traffic that is nonconforming with the SLA is the most likely to be dropped, and after that, priority goes according to the hierarchy of bandwidth types.

After receiving upstream resource requests from ONUs/ONTs, the OLT makes a bandwidth allocation decision in accordance with the QoS metrics and SLA. The ONU/ONT is responsible for intra-ONU scheduling, which means the ONU/ONT should arbitrate the transmission of different priority queues in its buffer. The OLT, for its part, performs inter-ONU scheduling to arbitrate the transmissions of different ONUs/ONTs. Information exchange between the OLT and an ONU/ONT is carried in PON MAC control messages. Figure 2.18 illustrates the major functional blocks supporting QoS in BPONs.

In IEEE 802.3 EPON, there is an MPCP for transmission arbitration that can be combined with priority traffic scheduling to realize QoS traffic priorities. It is reasonable to use the preferential scheduling mechanism of IEEE 802.1D [IEEE 802.1D] for this purpose. This was investigated by [KM1], who suggested schemes to eliminate a tendency to discriminate against lower-priority traffic under light loading. Chapter 3 provides more information on QoS mechanisms in EPONs.

2.3.3 Deployment and Maintenance

The deployment and maintenance of PONs requires care in layout and configuration, and adequate testing of subsystems and the completed network. Figure 2.19 [TELCORDIA] indicates testing points for a PON that includes a video broadcast capability.

Performance depends heavily on the transmission losses due to attenuation in the fiber and, to a lesser extent, on coupler and splice losses, and a loss budget must be adhered to. The maximum loss budget from OLT to the most distant ONU should be 25 dB for a class B PON, and 30 dB for a class C (longer distances, larger splitting ratio) PON, where ODN classes are defined in Table 2.1. Table 2.6 shows expected component losses and a typical specification of maximum acceptable transmission impairments.

In testing a new installation, the typical steps are [EXFO]

- total loss budget verification,
- link characterization of both the feeder and distribution links,

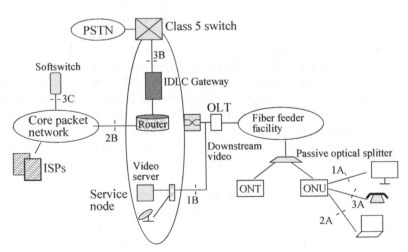

Figure 2.19 Test points in a PON carrying telephony, data, and video services. IDLC, integrated digital loop carrier [TELCORDIA].

TABLE 2.6 Typical Component Losses and Transmission Impairment Limits [EXFO]

Fiber loss: 0.33 dB/km at 1310 nm, 0.21 dB/km at 1490 nm, 0.19 dB/km at 1550 nm
Splitter loss: 16 dB for 1:32 splitter
Connector and splice losses: 2–3 dB from OLT to ONT
WDM coupler loss: 0.7–1.0 dB
OLT to most distant ONU maximum loss: 25–30 dB (see Section 2.3.3, para. 2)
Back reflection of coupler ports: –35 dB or below
Splice losses: 0.1 dB or smaller

- coupler and splice loss measurement,
- end-to-end loss and back reflection evaluation, and
- activation of OLT and ONT/ONUs.

These optical transmission tests should be augmented with system tests, for example, as described in [TELCORDIA]:

- Overall data transmission performance measurements (throughput, latency, and lost frames under various optical layer impairment and topological conditions)
- Discovery and ranging of ONUs from the OLT
- Bandwidth sharing (including DBA if implemented)
- QoS support (traffic conditioning, operation with diverse service classes and SLAs, DBA/QoS interaction, etc.)
- Protection mechanisms (if provided)
- Functioning of supported services (e.g., DS1 circuit emulation service, LAN emulation, voice over AAL1/AAL2, and voice and video quality)
- Conformance at external customer and network interfaces
- Operations, administration, maintenance, and provisioning (OAM&P), including configuration, fault, and security management)

2.3.4 Problems and Troubleshooting

The problems that often occur in a PON system include connector damage and misalignment, fiber cable bend and break, insufficient power level at ONUs/ONTs, and hardware problems in an OLT or ONU/ONT. Depending on the location of a problem, the affected number of subscribers on a PON may be one, several, or all. Troubleshooting identifies the location and the source of a particular problem. Generally speaking, if a problem occurs in the shared part, such as the OLT, feeder fiber, or splitter, all subscribers will be affected. If a problem occurs in the dedicated part, such as the distribution

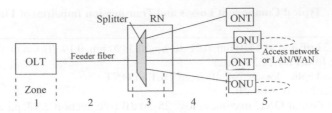

Figure 2.20 PON troubleshooting zones.

TABLE 2.7 Problem Causes and Locations

Problem	Cause	Location
One ONU/ ONT	▶Distribution fiber to the affected ONU/ONT	▶Zone 4
	▶Hardware failure of the affected ONU/ONT	▶Zone 5
Several ONUs/ ONTs	▶Power level is not enough for distant ONUs/ONTs	▶Zone 1
	▶Splitter output problem	▶Zone 3
All ONUs/ ONTs	▶No power at CO	▶Zone 1
	▶OLT failure	▶Zone 1
	▶Feeder fiber failure	▶Zone 2
	▶Splitter failure	▶Zone 3

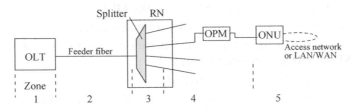

Figure 2.21 Troubleshooting using an optical power meter (OPM).

fiber or an ONU/ONT, the affected subscribers are essentially those downstream from the failed part.

In Figure 2.20, the PON system is segmented into five troubleshooting zones. Zones 1~3 are the shared parts, and zones 4 and 5 are the dedicated parts. The possible locations and causes of problems are tabulated in Table 2.7.

The types of troubleshooting equipment include optical power meters (OPMs) and optical time domain reflectometers (OTDR). An OPM measures the average power of a continuous light beam and consists of a solid-state detector, signal conditioning circuit, and digital display. It uses a removable adaptor for connection to tested devices. Unfortunately, an OPM is insensitive to rapid on–off fluctuations of bursty upstream traffic. Important specifications for OPMs include wavelength range and optical power range. In Figure 2.21, an OPM is connected to the distribution fiber to test the received optical power.

An OTDR measures the elapsed time and intensity of light reflected along an optical fiber. It is used for locating problems in a PON by computing the distance to breaks or attenuation. An OTDR injects a series of optical pulses into the fiber under test. It also extracts, from the same end of the fiber, light that is scattered or reflected back. The intensity of the returned pulses is measured over time and plotted as a function of fiber length, realizing a graph to locate fiber breaks. Additionally, the measured results can be used to estimate the fiber's length and overall attenuation. In Figure 2.22, an OTDR is connected to an ONT to locate a problem.

Figure 2.23 shows an example of an OTR trace. The power of the backscattered light, plotted versus distance (proportional to return time), is effectively

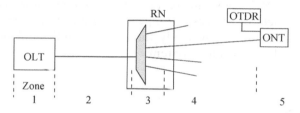

Figure 2.22 Troubleshooting using an optical time domain reflectometer (OTDR).

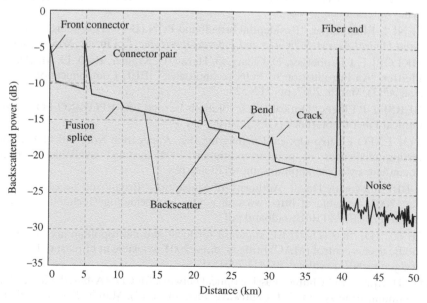

Figure 2.23 Example of an OTDR trace [RAD].

the impulse response of the link. The jumps in this example correspond to the IL of various components, and the large reflection peak suggests Fresnel reflections at the fiber termination.

Unfortunately, a PON will return traces from multiple branches, whose overlapping traces can make it very difficult to diagnose a problem in a particular branch. The many approaches to isolating traces from the different branches include [RAD]

- temporal isolation of branches, through the differing response times from different lengths of fiber to the different ONUs together with selective use of reflectors at each ONU; and
- use of one monitoring wavelength for each branch.

Other techniques are also possible, including the use of a unique optical coder just before each ONU, whose upstream signal, in a nondata wavelength band, identifies the branch. In any case, it appears possible, despite the multiplicity of branches to ONUs, to effectively monitor the individual branches in a central automated management system.

REFERENCES

[AGRAWAL] G. Agrawal, *Nonlinear Fiber Optics*, third ed., Elsevier, 2001.

[EDFA] E. Desurvire, *Erbium-Doped Fiber Amplifiers: Principles and Applications*, Wiley-Interscience, 1994.

[EFFEN] F. Effenberger, "Residential broadband PON (B-PON) systems," in *Broadband Optical Access Networks and Fiber-to-the-Home*, C. Lin, ed., Wiley, 2006.

[ECHKLOP] F. Effenberger, D. Cleary, O. Haran, G. Kramer, R.-D. Li, M. Oron, T. Pfeiffer, "An introduction to PON technologies," IEEE Communications Magazine, 45(3), March, 2007, pp. 517–525.

[ELBERS] J.-P. Elbers, "Optical access solutions beyond 10G-EPON/XG-PON," Proc. Optical Fiber Conference (OFC), 2010.

[EXFO] EXFO Electro-Optical Engineering Inc., Application Note 110, "Fiber-optic testing challenges in point-to-multipoint PON testing," available at http://documents.exfo.com/appnotes/anote110-ang.pdf

[IDA-BB] Infocomm Devel. Authority of Singapore, "Broadband access," November, 2002, available at http://www.ida.gov.sg/doc/Technology/Technology_Level1/20060417212727/ITR4Broadband.pdf

[IEEE 802.1D] "802.1D—IEEE standard for local and metropolitan area networks, media access control (MAC) bridges," June, 2004, available at http://standards.ieee.org/getieee802/download/802.1D-2004.pdf

[KIM] H. Kim & Y.C. Chung, "Passive optical network for CDMA-based microcellular communication systems," J. Lightwave Technol., 19(3), March, 2001, available at http://koasas.kaist.ac.kr/bitstream/10203/11983/1/ij55.pdf

[KM1] G. Kramer & B. Mukherjee, "Supporting differentiated classes of service in Ethernet passive optical networks," J. Optical Netw., 1(8), pp. 280–298, 2001.

[KM2] G. Kramer & B. Mukherjee, "Ethernet PON: Design and analysis of an optical access network," Photonic Netw. Comm. J., 3(3), July, 2001.

[RAD] M. Rad, K. Fouli, H. Fathallah, L. Rusch, & M. Maier, "Passive optical network monitoring: Challenges and requirements," IEEE Communications Magazine, February, 2011.

[TELCORDIA] "Passive optical network testing and consulting," available at http://www.telcordia.com/services/testing/integrated-access/pon/

[XG-PON] F. Effenberger, H. Mukai, S. Park, & T. Pfeiffer, "Next-generation PON-Part II: Candidate systems for next-generation PON," IEEE Commun. Magazine, November, 2009, pp. 50–57.

[YEH] C. Yeh, C.-S. Lee, & S. Chi, "A protection method for ring-type TDM-PONs against fiber fault," Prof. IEEE-OSA Optical Fiber Communication Conference 2007, paper JThA76.

3

TECHNIQUES AND STANDARDS

This chapter focuses on the three main passive optical network (PON) technologies and standards: broadband passive optical network (BPON), gigabit-capable passive optical network (GPON), and Ethernet passive optical network (EPON), which were introduced in the previous chapter. It explains the more significant characteristics of these three systems and provides an overview of the standards and recommendations, noting frame formats, data transmission functions, and operations, administration, and maintenance (OAM) functions and messages. Our focus, here as throughout the book, is on architecture, not on implementation details. Table 3.1 lists the major documents in the large set of recommendations and standards.

Organizations involved in PON standards development include the International Telecommunication Union-Telecommunications (ITU-T), the Institute of Electrical and Electronics Engineers (IEEE), the Internet Engineering Task Force (IETF), the International Electrotechnical Commission (IEC), the Telecommunications Industry Association (TIA), the American National Standards Institute (ANSI), the European Telecommunications Standards Institute (ETSI), and the Japanese Industrial Standards (JIS). ITU-T and IEEE publish standards and recommendations for the PON physical and medium access levels, while most of the others address issues related to the PON products. The IETF focuses on the network level and management, as exemplified by

The ComSoc Guide to Passive Optical Networks: Enhancing the Last Mile Access,
First Edition. Stephen Weinstein, Yuanqiu Luo, Ting Wang.
© 2012 Institute of Electrical and Electronics Engineers. Published 2012 by John Wiley &
Sons, Inc.

TABLE 3.1 PON Recommendations and Standards

Organization	PON Flavor	Recommendation/Standard	Issue Time
ITU-T	BPON	G.983.1: Broadband optical access systems based on passive optical networks (PON)	October 1998
		G.983.2: ONT management and control interface specification for BPON	June 2002
		G.983.3: A broadband optical access system with increased service capability by wavelength allocation	March 2001
		G.983.4: A broadband optical access system with increased service capability using dynamic bandwidth assignment	November 2001
		G.983.5: A broadband optical access system with enhanced survivability	January 2002
ITU-T	GPON and XG-PON	G.984.1: Gigabit-capable passive optical networks (GPONs): general characteristics	March 2003
		G.984.2: Gigabit-capable passive optical networks (GPONs): physical media dependent (PMD) layer specification	March 2003
		G.984.3: Gigabit-capable passive optical networks (GPONs): transmission and convergence layer specification	February 2004
		G.984.4: Gigabit-capable passive optical networks (GPONs): ONT management and control interface specification (OMCI)	June 2004
		G.984.5: Enhancement band for gigabit-capable optical access networks	September 2007
		G984.6: GPON reach extension	March 2008
		G.984.7: GPON long reach	July 2010
		G987: 10-gigabit-capable (XG-PON) systems: definitions, abbreviations, and acronyms	January 2010
		G987.1: XG-PON general requirements	January 2010
		G987.2: XG-PON physical media dependent (PMD) layer specification	January 2010
		G.987.3: XG-PON transmission convergence (TC) specifications	October 2010
		G.988: ONU management and control interface (OMCI) specification	October 2010
IEEE	EPON and 10G EPON	IEEE 802.3ah-2004 amendment: media access control parameters, physical layers, and management parameters for subscriber access networks	September 2004
		IEEE 802.3av: CSMA-CD access method and physical layer specifications Amendment 1: physical layer specifications and management parameters for 10-Gbps passive optical networks	September 2009

[RFC4836], "Definitions of Managed Objects for IEEE 802.3 Medium Attachment Units," April 2007. We discuss, in this chapter, the ITU-T and IEEE standards plus a brief overview of IETF Requests for Comments (RFCs) relevant to EPONs. We will use the notation optical network unit (ONU), optical network termination (ONT), and the combination, ONU/ONT, interchangeably, although, strictly speaking and as noted earlier, ONU refers to a unit on the user side of an optical access network that may still connect through an additional access network to the user's premises, while ONT is an ONU at the user's premises.

3.1 BPON OVERVIEW

The ITU-T G.983.x recommendation series specifies BPON, covering the management and control interface, wavelength allocation, bandwidth assignment, and survivability issues. BPON is a higher-performance extension of asynchronous transfer mode passive optical network (APON) that similarly employs ATM cells to encapsulate data transmitted between the OLT and ONUs/ONTs. It adds support for wavelength division multiplexing (WDM), dynamic and higher upstream bandwidth allocation, and survivability. Table 3.2 shows the line signal transmission rates supported by BPON, which uses a bipolar scrambled non-return-to-zero (NRZ) modulation, and Figure 3.1 exemplifies

TABLE 3.2 BPON Line Signal Transmission Rates

Downstream (Mbps)	Upstream (Mbps)
155.52	155.52
622.08	155.52
622.08	622.08
1244.18	155.52
1244.18	622.08

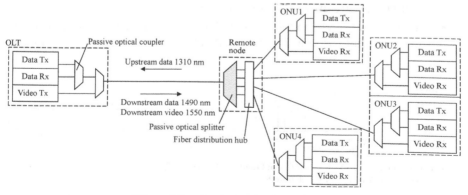

Figure 3.1 An example of a BPON.

a BPON supporting four ONUs. The combination of 622.08 Mbps downstream and 155.52 Mbps upstream for the line signaling rates is commonly accepted by the industry. The line signals consist of NRZ binary pulses, with the two levels being the "on" state (light being transmitted) and the "off" state (no light being transmitted). The line signaling rate must accommodate user information, framing overhead, and possible coding overhead.

3.1.1 Basic Asynchronous Transfer Mode (ATM) Concepts

A BPON transmission frame, in the usual implementation, consists of ATM cells and physical layer operation, administration, and maintenance (PLOAM) cells. ATM cells transport the information payload, while PLOAM cells carry grants for upstream transmission as well as other operations, administration, maintenance, and provisioning (OAM&P) messages. Upstream, similar PLOAM cells are used by ONUs to send their queue sizes to the OLT, which needs this information for its grant decisions.

ATM [McDysan] is a circuit-switching technique based on fixed-sized packets that are called cells. Its use of circuit switches, rather than routers, distinguishes ATM from Internet protocol (IP)-based networks. Although overshadowed by later developments, ATM is of historical importance for resource allocation in PONs and is reflected in flow reservation techniques, such as generalized multiprotocol label switching (GMPLS) of later optical systems. Figure 3.2 shows the structure of an ATM cell.

Each ATM cell is 53 bytes long, containing a 5-byte header and a 48-byte data payload. This length was a compromise between voice transmission needs for a short cell that could be filled with speech samples without introducing excessive delay, and data transmission needs for a long cell in which the header overhead would be relatively small. The following fields compose the header:

- Virtual path identifier (VPI), 8 or 12 bits, depending on whether the ATM cell is sent on the user–network interface (UNI) or the network–network interface (NNI). A VPI is effectively a "pipe" accommodating a multiplicity of virtual circuits (VCs) (or channels).
- Virtual circuit or channel identifier (VCI), 16 bits, designates a particular VC or channel carrying ATM cells. A VC or channel is a set of resource

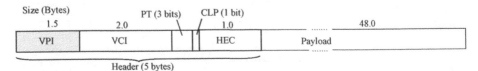

VPI: Virtual Path Identifier   VCI: Virtual Circuit Identifier   PT: Payload Type
HEC: Header Error Control   CLP: Cell Loss Priority

Figure 3.2 ATM 53-byte cell.

allocations at ATM switches that together define a connection through the network with specified quality-of-service (QoS) characteristics. VPI and VCI together compose the virtual path channel identifier (VPCI), indicating the connection information attached to the ATM cell.

- Payload type (PT), 3 bits, of which the first bit indicates whether the ATM cell contains user information or a control message. If the ATM cell contains user information, the second bit indicates congestion, and the third bit indicates whether the ATM cell is the last in a series of cells that represent a single ATM adaptation layer 5 (AAL5) frame. Various ATM adaptation layers are defined to support different information transfer needs. AAL5, devoted to packet data communications, segments data packets (such as IP packets) into cells at the transmitting terminal and reassembles the data packets at the receiving end.

- Cell loss priority (CLP), 1 bit, indicates whether the ATM cell should be discarded if it encounters extreme congestion.

- Header error control (HEC), 8 bits, is a checksum calculated from the ATM cell header only.

The downstream (OLT to ONU/ONT) PLOAM cell (Figure 3.3), also 53 bytes long, carries management messages, principally grants made by the OLT in a BPON system to end users. A grant is an upstream transmission allocation in a time division multiple access (TDMA) system, informing an ONU/ONT of the starting time and duration of its next upstream data burst. The following fields are contained in the downstream PLOAM cell:

- header (5 bytes), as in a regular ATM cell;
- IDENT (2 bytes), of which the first byte is set to 0 and the second byte is set to 1 if the PLOAM cell is the first within a downstream frame and set to 0 if it is not;
- synchronization (2 bytes), conveying a 1-kHz reference signal from the OLT to ONUs/ONTs to synchronize the downstream frame;
- grant (1 byte), one of up to 27 grants in one downstream PLOAM cell;
- cyclic redundancy check (CRC), 1 byte; there are totally 4 CRC bytes in one downstream PLOAM cell, each protecting a group of seven or six grants;
- message_field (10 bytes), to transport alarms and threshold-crossing alerts; and

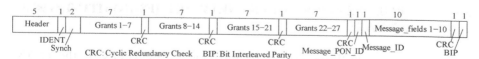

Figure 3.3 Downstream PLOAM cell.

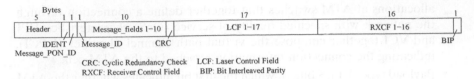

Figure 3.4 Upstream PLOAM cell.

- bit interleaved parity (BIP of all bits since the last BIP), 1 byte, for error detection on the downstream link.

The upstream PLOAM cell transfers the status of the ONU/ONT queue to the OLT (Figure 3.4). The major fields are

- header (5 bytes), as above;
- IDENT (1 byte, set to 0);
- message_field (10 bytes)
- CRC, 1 byte, protecting the message field;
- laser control field (LCF), 17 bytes, for physical layer use;
- receiver control field (RXCF), 16 bytes, for physical layer use; and
- BIP, 1 byte, for error detection on the upstream link.

One of ATM's strong points is its ability to effectively allocate capacity among different classes of traffic, making ATM networks useful for a variety of applications with different QoS requirements. The physical layer becomes the multiplexer for combining upstream traffic on a common fiber. The ATM classes are

- *Constant Bit Rate (CBR)*: The customer can transmit up to a specified bit rate. This supports steady-stream, low-latency traffic such as telephony and is analogous to traditional leased line service.
- *Variable Bit Rate (VBR)*: The customer contracts for specified average and peak rates, allowing moderate-cost service for fluctuating transmission speeds as in some video-intensive applications.
- *Available Bit Rate (ABR)*: The customer purchases a guaranteed minimum data rate and is allowed to burst at higher rates when uncommitted capacity is available.
- *Unspecified Bit Rate (UBR)*: No guaranteed rate, similar to "best effort" in IP networks; it may be suitable for applications that can tolerate delays.

3.2 THE FULL SERVICE ACCESS NETWORK (FSAN) (ITU-T G.983) BPON STANDARD

Section 2.1.3 succinctly describes the introduction of APON by the FSAN industry consortium and the basic capabilities of the BPON G.983 standard that followed. We provide additional detail here and in subsequent subsections on the standard and the main transmission techniques used to implement it. Ranging for synchronized upstream transmissions, used in all PON formats, is discussed under GPON in Section 3.3.4, and security is discussed in Section 3.3.5.

The ITU G.983 standards support the transmission rates shown in Table 3.2, as well as the multiple ATM service classes referred to in the previous section. As shown in Figure 3.1, data and video are transmitted downstream on the wavelengths of 1490 and 1550 nm, respectively, with data time division multiplexing (TDM), and data are transmitted upstream using TDMA on 1310 nm. The OLT uses a ranging technique to determine, and then to equalize, the different propagation times from the different end ONUs so that the TDMA system is properly synchronized among the OLT and all of the ONUs. Allocations of upstream capacity (slots in the TDMA system) are made using the downstream PLOAM cells described in the previous subsection. Any number of upstream slots can be allocated by the OLT to a given ONU, within the constraint that the total of all allocations does not exceed the total capacity.

The five BPON standards listed in Table 3.1, all available at http://www.itu.int/rec/T-REC-G/e, have the following functions [Sangha] [ITU-T]:

G.983.1 (1998): optical layer, transmission convergence (TC) layer, and ATM layer

G.983.2 (2000): operations protocol and message set (ONT management and control interface [OMCI]) between the OLT and the ONT

G.Imp983.2 (2005): Implementers' Guide for ITU-T Rec. G.983.2

G.983.3 (2001): alternative wavelength plan with an additional wavelength band for downstream video broadcast or for bidirectional transport of data using dense wavelength-division multiplexing (DWDM).

G.983.4 (2001): enhancement of service capability using dynamic bandwidth assignment

G.983.5 (2002): survivability-protection enhancements for the delivery of highly reliable services.

G.983.1 "describes a flexible optical fiber access network capable of supporting the bandwidth requirements of narrowband and broadband services [and] describes systems with nominal downstream line rates of 155.52, 622.08 and 1244.16 Mbps, and nominal upstream line rates of 155.52 and 622.08 Mbps. This Recommendation proposes the physical layer requirements and specification for the physical media dependent layer, the TC [transmission convergence] layer and the ranging protocol of an ATM-based Broadband Passive Optical Network (BPON)." Figure 3.5 shows the reference configuration, illustrating the notation for the connection interfaces. Note that the ONU is on the left and the OLT on the right, a reversal of the usual placement in diagrams

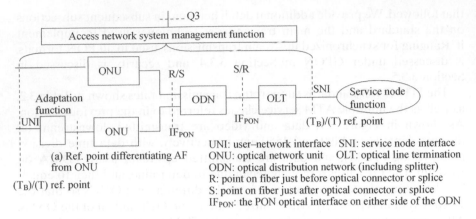

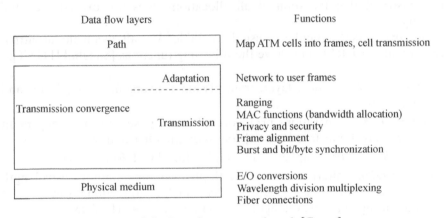

UNI: user–network interface   SNI: service node interface
ONU: optical network unit   OLT: optical line termination
ODN: optical distribution network (including splitter)
R: point on fiber just before optical connector or splice
S: point on fiber just after optical connector or splice
IF$_{PON}$: the PON optical interface on either side of the ODN

Figure 3.5   Reference configuration for an ATM-based BPON (G.983.1).

Data flow layers		Functions
Path		Map ATM cells into frames, cell transmission
Transmission convergence	Adaptation	Network to user frames
	Transmission	Ranging MAC functions (bandwidth allocation) Privacy and security Frame alignment Burst and bit/byte synchronization
Physical medium		E/O conversions Wavelength division multiplexing Fiber connections

Figure 3.6   Data flow protocol stack [Green].

in this book, and that the passive outside plant is called the optical distribution network (ODN).

The ONU and OLT together are "responsible for providing transparent ATM transport service between the UNI and the SNI." The protocol layers, shown in Figure 3.6, consist of the physical media dependent (PMD) layer (modulation systems for upstream and downstream channels), the TC layer (for managing distributed access to the upstream PON capacity across multiple ONUs), and the path layer (normal ATM functions including mapping cells into frames).

The TC layer in a BPON specifies many requirements not shown in detail in Figure 3.6, including cell rate decoupling (stuffing extra cells into a stream to reach a specified cell transmission rate), header error correction, the

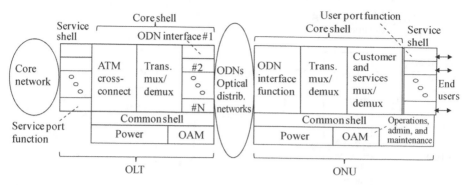

Figure 3.7 Functions within the BPON OLT and ONU. Note possible multiple ODNs.

maximum number of virtual paths (VPs) (4096), the minimum addressing capability (64 ONUs), and the transmission rates cited earlier. It specifies the payload capacities and the system for point-to-multipoint (P2MP) transmission, namely, downstream broadcast to all ONUs and upstream transmission from each ONU controlled by the OLT through grants for use of the TDMA upstream channel.

Figure 3.7 indicates the functional blocks within an ONU and an OLT. The OLT and ONU are both active with powered electronics. The interface in an ONU to the ODN, that is, the physical network carrying traffic between that ONU and the OLT, performs optoelectronic conversions, retrieves ATM cells from the payload in the downstream frame, and inserts ATM cells into the upstream payload, synchronized by the downstream frame timing. The multiplexer combines traffic from the customer service interfaces into the ODN (network side) interface.

The OLT, for its part, connects to the core network on one side through standard interfaces (VB5.x, V5.x, and other NNIs), while on the distribution side it offers optical access meeting service level agreements on bit rate, power, and other parameters. Its service port function, facing the core network, typically inserts ATM cells into the upstream synchronous optical network (SONET)/synchronous digital hierarchy (SDH) payload and retrieves ATM cells from the downstream payload. The multiplexer provides VP connections across the OLT, with different VPs assigned to different services at the IF$_{PON}$ interface. VCs within the VP carry user data, signaling, and management cells.

At the physical level, maintaining bit synchronization in downstream transmissions is an important requirement, and the following section is relevant for virtually all PONs, not only BPON. PONs use many of the same bit synchronization techniques employed in SONET and local area network (LAN) systems. Timing information is derived from transitions in the optical line signal, that is, from "high" to "low" and vice versa. Care must be taken to have a sufficient density of transitions, avoiding long periods of high or low signals. This is realized by scrambling in BPON and GPON, and block coding in

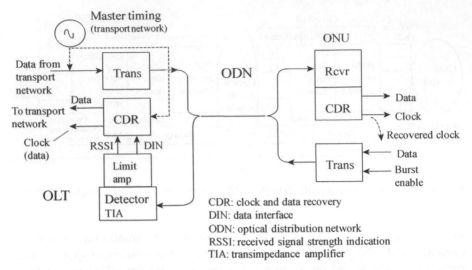

Figure 3.8   Loopback timing recovery in a PON.

EPON. With these measures, an ONU can use conventional low-cost optical receivers and clock and data recovery (CDR) circuits.

As illustrated in Figure 3.8, the OLT receives a stable clock from the transport network, and the ONUs are in turn locked to the timing master in the OLT. ONU receivers recover the clock frequency from the signal received from the OLT with the aid of filtering in a phase-locked loop (PLL) to remove jitter. For subsequent upstream signaling, the ONU sends symbols at a bit rate that is locked to the recovered downstream clocking frequency.

In contrast to traditional wide area network (WAN) or LAN links, the CDR in the OLT must, when multiple ONUs use the same upstream wavelength, be capable of rapidly recovering timing phase in the upstream bursts. This can be carried out by a fast-tracking PLL (analog phase aligner), a gated oscillator, or by multiphase oversampling with digital phase alignment, all illustrated in Figure 3.9. The preamble in a transmission frame provides a training signal long enough for reliable timing phase lock. The OLT's receiver must also tolerate the gaps in transmission associated with guard bands.

At the medium access protocol level, the downstream and upstream frame structures are described in the following two subsections, relying on G.983.1 and [Green].

3.2.1   Downstream Transmission

Figure 3.10 illustrates the 125-ms downstream frame format for 155.52-Mbps BPON transmission from the OLT to the ONUs/ONTs. The contents are proportionately higher for higher transmissions rates, as described below. We have

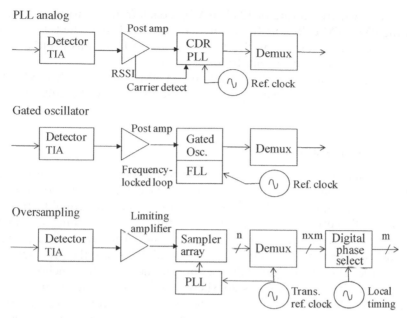

Figure 3.9 Bit timing phase alignment using a PLL, a gated oscillator, or oversampling.

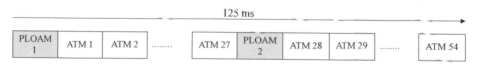

Figure 3.10 Downstream frame for 155 Mbps.

already seen, in Figure 3.3, the fields within a PLOAM cell for assignment of upstream transmission grants. A 1-byte grant informs a particular ONU whether it can transmit, in the next upstream frame, either a certain number of ATM data cells in specified time slots or a (single) PLOAM cell.

For the downstream rate of 155.52 Mbps, each frame consists of 54 ATM cells and 2 downstream PLOAM cells, each in a 53-byte downstream slot. The total number of upstream transmission grants (of 56-byte upstream slots) carried within a downstream BPON frame is 53, of which 27 grants are indicated in the first downstream PLOAM cell and 26 in the second.

For the downstream rate of 622.08 Mbps, each frame consists of 224 cells, including 216 ATM cells and 8 downstream PLOAM cells, again separated by 27 ATM cells. The total number of upstream slot grants carried by the PLOAM cells in a downstream BPON frame is either 53 or 212, depending on the upstream transmission rate.

For the downstream rate of 1244.16 Mbps, each frame consists of 448 cells, including 432 ATM cells and 16 downstream PLOAM cells. The total number of upstream grants carried by the downstream PLOAM cells is either 53 or 212, depending on the upstream transmission rate.

Downstream frames are broadcast in a TDM system. Each ONU/ONT receives all of the data, passing to its subscribers only the cells destined for them.

3.2.2 Upstream Transmission

Upstream transmission from ONUs/ONTs to the OLT is in the form of cell bursts, either ATM cells or upstream PLOAM cells. A 3-byte overhead is inserted in front of each cell, making the required slot size equal to 56 bytes. The ATM cell is effectively encapsulated within the system time slot. Figure 3.11 illustrates the contents of the upstream slot. The guard time provides 1-byte delay between consecutive cells to avoid collisions; the preamble supports synchronization and gain adjustment; and the delimiter synchronizes the slot beginning. These are well-established techniques for synchronization.

Figure 3.12 shows the upstream frame structure, illustrating both fixed slot assignments (a particular ONU gets the same from frame to frame) and assign-

Figure 3.11 Contents of a 57-byte BPON upstream slot, with three prefix bytes to support ODN physical-layer functions.

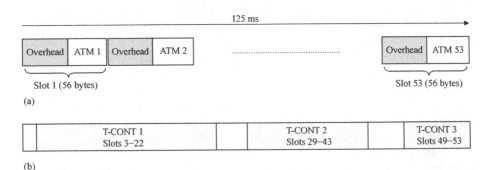

Figure 3.12 Upstream frame structure examples. (a) Individually assigned ATM cells in a 155 Mbps frame. (b) Multiple cell (T-CONT) grants to particular ONUs, with blank spaces representing individually assigned ATM cells.

ment of bursts of slots for customer traffic, called transmission containers (T-CONTs), implementing dynamic bandwidth assignment (dynamic bandwidth allocation [DBA]), responding to ONU transmission needs. A T-CONT is a bundle of upstream traffic flows from the same ONU/ONT with the same characteristics. The BPON with DBA has much greater and faster flexibility in sharing the upstream capacity among the ONUs in accord with their dynamically changing requirements.

For the upstream rate of 155.52 Mbps, shown in Figure 3.12a, each frame consists of 53 time slots, and each time slot carries 56 bytes (a 3-byte overhead plus a 53-byte ATM or PLOAM cell). For the upstream rate of 622.08 Mbps, which is not shown in Figure 3.12, each frame consists of 212 time slots, and each time slot carries 56 bytes (a 3-byte OH plus a 53-byte ATM or PLOAM cell).

3.2.3 Management Functions

G.983.2, the ONT (the same as our ONU) management and control interface, "addresses the ONT configuration management, fault management, and performance management for BPON system operation" and for a number of specified services including ATM adaptation layers 1, 2, and 5; circuit emulation service; Ethernet services; IP routing; wireless LAN service; asymmetric digital subscriber line (ADSL) and very high-speed digital subscriber line (VDSL) services; voice services; WDM; PON protection switching; dynamic bandwidth assignment (DBA); and security. The managed entities are abstract representations of resources and services in an ONT/ONU.

A comprehensive discussion of the management functions is beyond the scope of this handbook, but it is helpful to consider one example, management of the upstream DBA function, within the ONU, described earlier as implemented with T-CONT cell bursts. Two models are supported by the standard, the first, where priority queues, traffic schedulers, and the T-CONT buffers are related in a fixed fashion, and the second, where these elements can be associated in a flexible way. We illustrate in Figure 3.13 the first model, showing the entities configured by management messages. Traffic schedulers can be added within a T-CONT buffer, providing differentiated service levels to individual priority queues or combinations of priority queues within that buffer.

3.2.4 Wavelength Division Multiplexing (WDM)

G.983 "defines new wavelength allocations to distribute ATM-PON signals and additional services signals simultaneously. New wavelength bands for additional services are made available by constraining the current ATM-PON downstream wavelength to a portion of downstream optical spectrum originally specified in ITU-T G.983.1." In essence, this standard introduces WDM

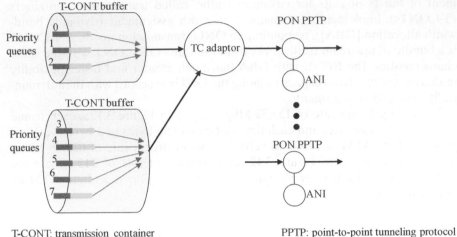

T-CONT: transmission container
TC adaptor: transmission convergence adaptor
(provides traffic to working or protection line termination)

PPTP: point-to-point tunneling protocol
ANI: access node interface

Figure 3.13 Management model for a flexible set of priority queues, traffic schedulers, and T-CONT buffers.

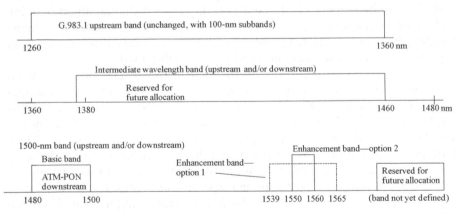

Figure 3.14 Wavelength allocations in G.983.3. Option 1 in the enhancement band is for additional digital services, while option 2 is for video distribution services.

to enhance the service capabilities of the earlier G.983.1 system. The greatest benefit may be simply multiplying the capacity of the installed fiber plant by upgrading only at fiber termination points. Figure 3.14 shows the wavelength allocations specified in the standard.

Note that the enhancement band "may be used not only for downstream but also for upstream signal transmission in the WDM scheme." The enhancement band coincides with the optical C-band, where it is possible to employ an erbium-doped fiber amplifer (EDFA) for power boosting.

3.2.5 Dynamic Bandwidth Allocation (DBA)

G.983.4 specifies the DBA method already introduced above as a function within the TC layer. DBA transforms a system that previously was appropriate for steady streams of traffic into one more capable of accommodating bursts of activity. G.983.4 specifies two DBA mechanisms: "idle cell adjustment" and "buffer status reporting (SR)."

In the idle cell adjustment mechanism, the OLT keeps track of the bandwidth usage of each ONU/ONT by observing the number of idle cells in the upstream frame. The OLT assigns more bandwidth to an ONU/ONT when the received traffic is heavy and there are few idle cells in the upstream time slot. Since the ONUs/ONTs play a passive role without sending their status, this mechanism is also called the nonstatus reporting (NSR) strategy.

In the buffer SR mechanism, the ONUs/ONTs actively report their buffer status to the OLT. The buffer status report indicates the instantaneous bandwidth requirement of an ONU/ONT, enabling the OLT to quickly determine the upstream bandwidth allocation. This mechanism is also referred to as SR strategy.

3.2.6 Protection Switching

G.983.5 "describes BPON survivability architectures, protection performance criteria, and protection-switching criteria and protocols." Some customers will justify greater expenditure of resources to achieve higher reliability. G.983.5 explains protection features and specifications, including "guidelines on performance objectives (e.g., switching and detection time), application functionality (e.g., revertive mode, nonrevertive mode, extra traffic support, automatic switching, and forced switching), switching criteria and switching protocols (1 + 1, 1:1, 1:N, bidirectional and unidirectional mechanisms)." It addresses the protection-configuration types B and C described in Appendix IV of G.983.1 and introduced below.

Configuration types B and C refer, respectively, to protecting the OLT only and protecting both the OLT and the ONUs. Figure 3.15 illustrates these two types, with duplicated units indexed by 0 and 1. If an initially activated unit fails, its functions are taken up by the redundant unit. In the "revertive" mode, functionality switches back to the original unit when it is repaired, while in the nonrevertive mode, the redundant unit remains in operation and the original unit becomes the backup.

Protection may be provided in various protocols:

1:1 Protection: The working unit carries all traffic, while the backup unit stands in reserve without serving any traffic.

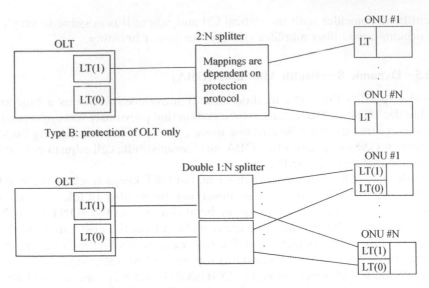

Type B: protection of OLT only

Type C: protection of both OLT and ONU, protocol either 1:1 or 1 + 1

Figure 3.15 Configuration types B (protection for OLT only) and C (protection for both OLT and ONUs). "LT" is an ODN interface as shown in Figure 3.7. The interface to the core network is typically in the V5.x series.

1 + 1 Protection: Identical traffic is carried in both working and backup equipment.

X:N Protection: X backup (protection) PONs are associated with N working PONs, with some or all of the protected ONUs on the working PONs connected to some or all of the X protection PONs.

Many further details of protection switching, including mixing of protected and unprotected facilities, multivendor interoperability between OLT and ONU, the location of the protection function within the layered model (PON physical, section, protection, and VP), and the role of PLOAM cells are offered in the standard.

3.3 GPON

GPON, defined in the G.984 family of standards (Table 3.1), is an enhancement of BPON, inheriting most of the characteristics described above. It supports higher rates, better security, and a choice among ATM, GPON encapsulation method (GEM) and Ethernet transmission protocols rather than the single ATM protocol of BPON. GPON significantly increases bandwidth and bandwidth efficiency by supporting large, variable-length packets characterizing

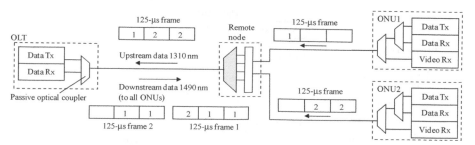

Figure 3.16 An example of GPON, showing TDMA multiplexing of upstream frames.

TABLE 3.3 GPON and XG-PON1 Transmission Rates

Downstream (Gbps)	Upstream
1.24416	155.52, 622.08, 1244.16 Mbps
2.48832	155.52, 622.08, 1244.16, 248,832 Mbps
9.95328	2.48832 Gbps

much of the traffic on IP networks. The aggregate downstream and upstream transmission rates range up to 2.488 Mbps. GEM efficiently aggregates user traffic and includes frame segmentation (guaranteed allocation of part of a frame) to reduce latency and thus to improve QoS for delay-sensitive voice and video traffic.

The G.984 standards specify the PMD layer, the TC layer, and ONU/ONT control interface. Figure 3.16 illustrates a GPON consisting of two ONUs. As with BPON, the downstream transmission is broadcast, while the upstream transmission is TDMA. The supported transmission rates over GPON are listed in Table 3.3.

3.3.1 GPON Encapsulation Method (GEM)

In order to carry traffic with variable packet sizes, the GEM encapsulates variable-length packets while minimizing the encapsulation overhead (which is 5 bytes per frame [OLA]). In particular, GEM carries TDM and Ethernet traffic over the GPON network using a variant of the generic framing procedure (GFP), an efficient encapsulation model defined by ITU-T Recommendation G.7041.

Figure 3.17 defines a GEM header and illustrates how Ethernet and TDM data are mapped into a GEM frame. The GEM frame has a 5-byte header that delineates the start of a GEM frame, allowing an OLT to identify the upstream frames (although the physical control block downstream (PCBd), described later, provides the frame's bit synchronization) without monitoring the

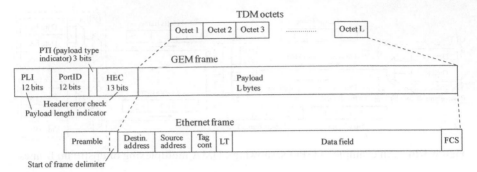

Figure 3.17 Mapping the alternatives of TDM octets and Ethernet frames into a GEM frame.

alignment information between consecutive frames. The header has the following fields:

- *Payload Length Indicator (PLI)*: Its value L is the number of bytes in the payload part of the GEM frame.
- *Port ID*: identifies the GFP port a GEM frame belong to.
- *Payload Type Indicator (PTI)*: indicates whether the GEM frame carries user data or a GEM OAM message. If it carries user data, the PTI field also indicates whether the GEM frame is the end of a data frame.
- *Header Error Check (HEC)*: error detection and correction for the GEM frame header.

GEM maps an Ethernet frame directly into the payload field. As illustrated in Figure 3.15, the fields in an Ethernet frame, including destination address (DA), source address (SA), length/type (LT), operation code (Opcode), time stamp (TS), Ethernet payload, and frame check sequence (FCS), are mapped octet by octet into the GEM payload. For TDM data, GEM accumulates a group of consecutive TDM data values and similarly maps them into the GEM payload octet by octet.

By fragmenting data at the sender into GEM frames and reassembling them at the receiver, GEM provides an appropriate encapsulation layer over GPON, and no other encapsulation layer is required. This facilitates a dramatic increase of bandwidth provisioning.

3.3.2 Downstream Transmission

GPON inherits the OAM mechanism from BPON by taking over the PLOAM message and major features of the OMCI. The GPON transmission convergence (GTC) layer maps data, including GEM frames, onto the GPON physical layer by using downstream and upstream frames. As in BPON, each GPON

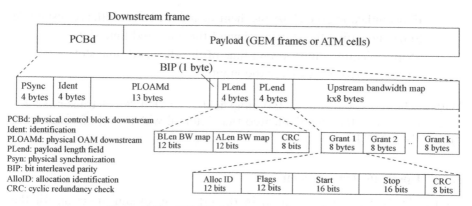

Figure 3.18 GPON downstream frame format.

frame is of 125-μs duration, with the number of bytes in a frame depending on the transmission rate. At 1.24416 Gbps, one frame contains 19,440 bytes, while at 2.48832 Gbps, one frame contains 38,880 bytes, with proportionally higher contents for next-generation passive optical network (XG-PON).

Figure 3.18 shows the format of the GPON downstream frame, beginning with a PCBd. The payload follows the PCBd, and it can be ATM cells, GEM frames, or a mixture of the two. In the case of mixed payload, the ATM cells immediately follow the PCBd, and the total number of ATM cells is specified by the ALen field in the PCBd. Following the ATM cells, the GEM frames occupy the remainder of the payload.

The fields in the PCBd, which can be monitored in a central office or other control points, are

- physical synchronization (Psync), indicating the beginning of the downstream frame;
- identification (Ident), indicating whether forward error correction (FEC) is used, and containing the superframe count;
- physical layer operation, administration, and maintenance downstream (PLOAMd);
- BIP, providing error detection for the downstream frame;
- PLend (payload length field), consisting of a BLen field for the number of grants carried by the downstream frame, the ALen field for the number of ATM cells after the PCBd, and a CR
- C for error correction; the Plend field is sent twice for robustness;
- upstream bandwidth (BW) map (upstream BW grants), consisting of k 8-byte grants, each grant carrying the following fields:
  - Allocation identification (AllocID), indicating to which ONU the bandwidth allocation applies

○ Flags, indicating the DBA mechanism and the bandwidth request mode
○ Start, indicating the starting time of the allocated time slot
○ Stop, indicating the stopping time of the allocated time slot
○ CRC, an error check over the grant.

With the help of the upstream BW map field, each ONU/ONT transmits buffered data to the OLT in a dedicated time interval without interference from other ONUs/ONTs.

3.3.3 Upstream Transmission

As in BPON, T-CONTs are used in the GPON upstream direction for the management of the customer data that they transmit. A GPON T-CONT can either carry ATM cells or GEM frames, but not both simultaneously. One ONU/ONT may support multiple T-CONTs, with each T-CONT being identified by its unique AllocID in the GPON network. The OLT arbitrates the upstream bandwidth based on the AllocIDs.

Figure 3.19 shows the upstream frame, with the following fields:

- physical layer overhead upstream (PLOu), indicating the beginning of the upstream frame, the BIP information, and the ONU ID;
- physical layer operation, administration, and maintenance upstream (PLOAMu), carrying an upstream PLOAM message;
- power leveling sequence (PLS), for power control measurements by the ONU; and
- dynamic bandwidth report upstream (DBRu), indicating the T-CONT traffic status.

Because one ONU/ONT may support multiple T-CONTs, there can be multiple "DBRu" and "payload" fields in one ONU/ONT's upstream burst. As illustrated in Figure 3.19, the burst from ONU1 carries subtime slots from two

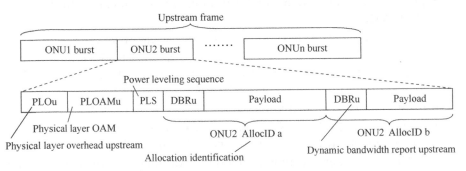

Figure 3.19 GPON upstream frame format.

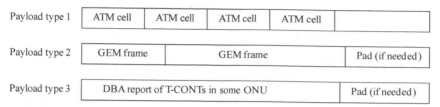

Figure 3.20 GPON upstream payload types, in some ONU burst (as in Figure 3.19).

T-CONTs (AllocID a and AllocID b). The payload falls in one of the three types: ATM cells, GEM frames, or ONU DBA report message. Figure 3.20 shows the upstream PTs. The payload of the ONU DBA report (PT 3 in Figure 3.20) contains a group of DBA reports of the T-CONTs from the same ONU.

G.984.3 provides a priority capability of transmitting different GEM frames with different QoS guarantees. This is done by assigning GEM frames to different T-CONTs with different levels of bandwidth guarantee. There are five such levels for T-CONTs:

Type 1: fixed bandwidth
Type 2: assumed bandwidth
Type 3: assured, nonassured, and maximum bandwidth
Type 4: best effort
Type 5: any type of bandwidth assurance, whatever is available

The port ID field of a GEM frame indicates what T-CONT it is in and thus the appropriate QoS treatment. However, the transparency of GEM encapsulation results in the GEM frame carrying no indication of the priorities of, for example, priority-labeled Ethernet packets that are encapsulated in the frame, so that it is not obvious how the GEM frame can be assigned to the appropriate priority-level T-CONT. This problem can be solved by, for example, insertion of precedence information, derived from the priority levels of the encapsulated packets, into the port identifier field of the GEM frame [Capurso].

3.3.4 Ranging

Since ONUs/ONTs may be located at different physical distances from the OLT, ranging is necessary to coordinate the upstream transmissions so that they arrive in the correct time slots as seen by the OLT. Ranging avoids upstream transmission collisions by determining the specific transmission delay from each ONU/ONT.

In GPON, the OLT initiates the ranging process by sending a ranging request message downstream to a particular ONU/ONT, which replies within a ranging window (a practical limit on the variation in propagation time) in the upstream direction. The OLT then calculates the equalization delay based

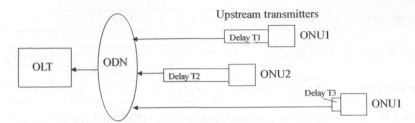

Figure 3.21 Compensating delays in upstream transmission determined by the ranging process.

on the round-trip-delay (RTD) time collected from the ranging window and feeds this information back to the ONU/ONT. The equalization delay is employed by the ONU/ONT, as shown in Figure 3.21, as the offset between the start of its received downstream frame and the start of its upstream frame. As a result, all ONUs/ONTs appear to be placed at the same virtual distance from the OLT, and their upstream transmissions arrive at the OLT without conflict. Ranging maintains the range knowledge at the OLT to high accuracy, with an uncertainty of only 8 bits at 1.2 Gbps [GREEN] and thus a minimal guard space.

3.3.5   Security

Downstream data in GPON, as in BPON, are broadcast to all ONUs/ONTs via the shared feeder fiber, and each ONU/ONT receives all of the downstream data. In order to ensure privacy, GPON, as in BPON, specifies an encryption process to secure downstream transmissions.

Specifically, the OLT launches the key exchange process by sending a PLOAM message to the ONU/ONT. The ONU/ONT generates a key and feeds it back to the OLT, relying on the high level of optical isolation in upstream transmissions to keep this key a secret from the other ONUs. After that, the OLT employs the Advanced Encryption Standard (AES) mechanism to encipher the downstream data on a 16-byte block basis using that key. Among all ONUs/ONTs, only the one that supplied the key can correctly decipher the encrypted downstream data. BPON initially relied on "churning" (change of a private encryption key at least once per second) for enhanced downstream security but later added AES, with its less frequent key changes, as an alternative.

3.4   EPON

The standardization of EPON was motivated by the traditional advantages of Ethernet. Ratified in 1983 as the IEEE 802.3 standard, Ethernet has been ubiquitously deployed in networks. It dominates LANs, and high-speed switched Carrier Ethernet is becoming ubiquitous in metropolitan area networks (MANs) and WANs as well. The major features of Ethernet include the following:

- low-cost Ethernet UNI,
- unique medium access control (MAC) address for low-cost layer 2 Ethernet switching,
- designed to carry traffic with variable packet sizes, and
- it is the dominant LAN technology.

EPON was proposed by the IEEE 802.3ah Ethernet in the First Mile (EFM) Task Force. The standard was ratified as IEEE 802.3ah-2004 in September 2004. EPON adopts Ethernet frames as the transmission data units and provides a symmetric line rate of 1.250 Gbps. An 8B/10B line encoding, developed by IBM in the early 1980s (http://i-data-recovery.com/data-recovery-service/hard-disk/what-is-sata-hard-disk), results in a user data rate of 1 Gbps. The design of the 8B/10B encoding ensures that there are never more than four 0s (or 1s) transmitted consecutively, and also that there are never more than six or fewer than four 0s (or 1s) in a single encoded 10-bit character, thus guaranteeing frequent level transitions for timing recovery. Longer encodings, such as 64B/66B, offered in 10G Ethernet, provide lower overhead. EPON uses a downstream wavelength between 1480 and 1580 nm, and an upstream wavelength between 1260 and 1360 nm.

The IEEE 802.3av standard for an enhanced version of EPON, the 10-Gbps EPON (10G EPON), was issued in late 2009. The user data rates are 10 Gbps downstream (from a 10.3125-Gbps line rate with 64B/66B code) and, in the upstream direction, either 1 Gbps (from a 1.25-Gbps line rate with 8B/10B code) or 10 Gbps (from a 10.3125-Gbps line rate with 64B/66B code). 10G EPON makes several advances over 1G EPON in transmission and energy efficiency and reliability [ANSARI]. As shown in Table 3.4 [TANAKA], it

- replaces the 8B/10B line coding of 1G EPON with 64B/66B line coding, reducing the overhead to as little as 3%;
- specifies mandatory FEC, using a Reed–Solomon (255, 223) code, while 1GPON has an optional (255, 239) Reed–Solomon code; and
- has an extended group of power budget classes for asymmetric (10 Gbps downstream and 1 Gbps upstream) and symmetric (10 Gbps in each direction) operational modes, in particular, PR(X)30, supporting a 32-line split over at least 20 km.

However, much remains the same, in part to support backward compatibility. The EPON frame format, MAC data and control layers, higher-layer functions, OAM, and DBA are essentially the same as in 1G EPON.

TABLE 3.4 [TANAKA] Comparison of 1G EPON and 10G EPON

	Data Rate (Gbps)		Line Coding	FEC	Tx Type and Launch Power (dbm)		
	Upstream	Downstream					
1G-EPON	1.25	1.25	8b/10b	Optional RS(255, 239)	PX10: OLT: (−3,2) ONU: (−1,4)	PX20: OLT: (2,7) ONU: (−1,4)	
10G-EPON	10.3125	10.3125	64b/66b	Enabled RS(255, 223)	PR10: OLT: EML (1,4) ONU: DML (−1,4)	PR20: OLT: EML+AMP (5,9) ONU: DML (−1,4)	PR30: OLT: EML (2,5) ONU: HP DML (4,9)
	1.25	10.3125	64b/66b	Enabled RS(255, 223)	PRX10: OLT: EML (1,4) ONU: DML (−1,4)	PRX20: OLT: EML+AMP (5,9) ONU: DML (−1,4)	PRX30: OLT: EML (2,5) ONU: DML (6,5.6)

10G EPON was motivated by the market drive for delivery of digital video to subscribers. The 10-Gbps data rate would permit delivery of multiple high-definition video streams to each ONU/ONT with a split ratio of 1:32 or even higher. 10G EPON is applicable in multiple environments to support bandwidth-intensive applications that will require fast, reliable, scalable, first-mile connections. The targeted applications include broadcast TV, Internet protocol television (IPTV), video on demand (VoD), 3-D online interactive games, ultra-high-speed Internet, personal videocasting, business Ethernet access, distributed network attached storage, and medical imaging. The 10G interfaces are expected to eventually exhibit a ratio of interface to end-appliance cost comparable to that of the original EPON.

3.4.1 EPON Switched Ethernet

Unlike BPON and GPON, of which the latter encapsulates Ethernet frames within GEM frames, the actual data units transferred through EPON are Ethernet frames. As shown in Figure 3.22, there are no fixed upstream and downstream TDM frames.

EPON relies on switched Ethernet, the basis now for almost all Ethernets. Ethernet packets (also called frames) from different ONUs are transmitted without interference, with the help of buffers in the ONUs, to the OLT, where they are queued and switched into the backbone network, rather than occasionally colliding and being retransmitted as in the original contention bus Ethernet that implemented carrier sense multiple access—collision detection (CSMA-CD). Furthermore, multiple queues for different priority services may be maintained at the switching point, with Ethernet frames carrying a priority field when it is needed, as explained below.

Figure 3.23 shows the structure of IEEE 802.3 Ethernet frames. The DA and SA identify the receiver and sender of the Ethernet frame, respectively. Every Ethernet device contains a unique 48-bit MAC address assigned by the manufacturer. The payload length of an Ethernet frame ranges from 40 to 1494 bytes, making it suitable to carry traffic with variable packet size. Note the optional tag control field in the header, which carries a 3-bit priority value that selects one of eight queues of ordered priority in the next Ethernet switch,

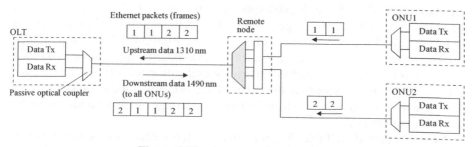

Figure 3.22 An example of EPON.

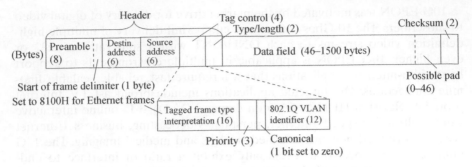

Figure 3.23   Ethernet frame, including priority field.

TABLE 3.5   Features of 1000BASE-PX10 and 1000BASE-PX20

	1000BASE-PX10		1000BASE-PX20	
Description	Upstream	Downstream	Upstream	Downstream
Data rate (Gbps)	1	1	1	1
Wavelength (nm)	1310	1490	1310	1490
Available power budget (dB)	23	21	26	26
Channel insertion loss (min/max, dB)	5/20	5/19.5	10/24	10/23.5
Allocation for penalties (dB)	3	1.5	2	2.5
Return loss (dB)	20	20	20	20
Max split ratio	1:16 (1:32 with FEC)		1:16 (1:32 with FEC)	
Physical distance (km)	10		20	

which will be the OLT for the case of upstream transmissions. The virtual LAN (VLAN) value identifies a virtual network associating a subset of communicating entities, such as those participating in a particular communication session.

3.4.2   1000BASE-PX10, 1000BASE-PX20, and 10G EPON PMD Types

The IEEE standard 802.3ah-2004 specifies two types of EPON systems, with different geographical coverage. 1000BASE-PX10, with a maximum reach of 10 km, is the IEEE 802.3 physical layer specification for a 1000-Mbps P2MP link over a single mode optical fiber. 1000BASE-PX20, with a maximum reach of 20 km, is the IEEE 802.3 physical layer specification for a 1000-Mbps P2MP link over a single mode optical fiber. The IEEE 802.3av standard specifies 10G EPON, with a maximum reach of 20 km and stronger FEC in addition to its much higher speed.

As summarized in Table 3.5, the three systems share the same network topology and management protocol, and 1310- and 1490-nm wavelengths are employed for upstream and downstream transmissions, respectively. Their maximum split ratio is 1:16 without FEC and 1:32 with FEC. The major difference relies on the physical specifications such as channel insertion loss, power budget, and maximum distance 10G EPON features both asymmetric and symmetric rates. The rates of an asymmetric 10G EPON are 10 Gbps downstream and 1 Gbps upstream. A symmetric 10G EPON system provides 10-Gbps rate in both downstream and upstream. PMD types of 10G EPON are listed in Table 3.6.

3.4.3 Medium Access Control (MAC)

EPON relies on an asymmetric P2MP architecture to provide broadband access. In the downstream transmission, data are broadcast from the OLT to each ONU/ONT using the entire bandwidth of the downstream channel at 1490-nm wavelength. ONUs/ONTs selectively receive frames destined to them by matching the DAs encapsulated in the Ethernet frames. Multicast addressing is possible, supporting downstream multimedia services such as video broadcasting.

In the upstream direction carried on the 1310-nm wavelength, multiple ONUs/ONTs share a common upstream channel, as in B-PON and GPON. Only a single ONU/ONT may transmit during a time slot in order to avoid data collisions. The MAC control mechanism, including ranging as described in an earlier section, schedules upstream transmissions and makes efficient use of the EPON system. The multipoint control protocol (MPCP) is defined to support the effective transmission between one OLT and multiple ONUs/ONTs, as described in the following paragraphs. A more detailed description of the EPON MAC is presented in [GREEN].

We first consider the MAC control messages. MPCP is a frame-based protocol employing five MAC control messages, specifically REGISTER_REQ, REGISTER, REGISTER_ACK, GATE, and REPORT. Figure 3.24 illustrates the REGISTER_REQ, REGISTER, and REGISTER_ACK messages associating an ONU/ONT with an OLT for purposes of MAC.

The REGISTER_REQ message is a 64-byte Ethernet frame, generated by an ONU/ONT. The message is identified in the Opcode field. The flag field indicates special requirements for the registration (e.g, register to or deregister from an EPON system). The pending grant field states the maximum number of grants the ONU can buffer.

The REGISTER message is generated by an OLT to reply to the registration request from an ONU/ONT. Finally, the REGISTER_ACK message is generated by the ONU/ONT to echo back the registration parameters to the OLT.

The control messages for bandwidth assignment are GATE and REPORT, shown in Figure 3.25. An OLT broadcasts a GATE message downstream to its ONUs/ONTs, indicating its upstream bandwidth allocations. Each GATE message can carry up to four grants. In addition to bandwidth allocation, the

TABLE 3.6 Features Of 10G EPON PMD Types

Description	PR10 US (Upstream)	PR10 DS (Downstream)	PR20 US	PR20 DS	PR30 US	PR30 DS	PRX10 US	PRX10 DS	PRX20 US	PRX20 DS	PRX30 US	PRX30 DS
Data rate (Gbps)	10	10	10	10	10	10	1	10	1	10	1	10
Wavelength (nm)	1270	1577	1270	1577	1270	1577	1310	1577	1310	1577	1310	1577
Available power budget (dB)	23	21.5	27	25.5	32	30.5	23	21.5	26	25.5	30.4	30.5
Channel insertion loss (min/max, dB)	5/20		10/24		15/29		5/20		10/24		15/29	
Allocation for penalties (dB)	3	2.5	3	1.5	3	1.5	3	2.5	2	1.5	1.4	1.5
Return loss (dB)	20		20		10		20					
Physical distance (km)	10		20		10		10					

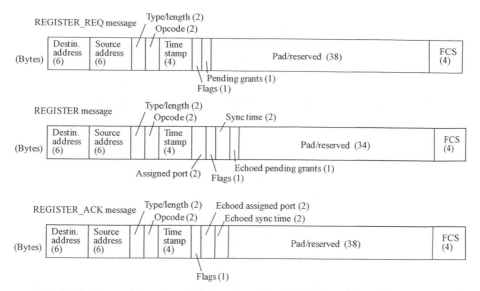

Figure 3.24   REGISTER_REQ, REGISTER, and REGISTER_ACK messages.

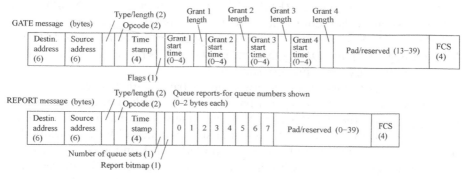

Figure 3.25   GATE and REPORT messages.

GATE message can be used to discover a new ONU/ONT, as will be illustrated in the next subsection. As Figure 3.23 shows, an ONU/ONT sends a REPORT message to the OLT, reporting its queue status. One REPORT message may carry up to eight queue statuses to the OLT.

The previous paragraphs described the MAC control messages that register ONU/ONTs and manipulate the medium access, assuming a set of ONU/ONTs known to the OLT. When a new ONU/ONT is attached, a discovery process, illustrated in Figure 3.26, facilitates registering it.

In the downstream transmission, the OLT periodically broadcasts a discovery gate message, which indicates the starting time and length of the special Discovery Time Window for registration. After receiving this message, a new

Figure 3.26   Registration of an ONU/ONT in the discovery process.

Figure 3.27   EPON bandwidth negotiation.

ONU/ONT waits for the beginning of the Discovery Time Window and then transmits a REGISTER_REQ message, which contains the ONU/ONT's MAC address and other specifications, to the OLT. The OLT then registers the ONU/ONT, replying with a REGISTER message back to the ONU/ONT. A GATE message is also sent to the ONU/ONT to assign a time slot to transmit a REGISTER_ACK message. Upon successful receipt of the REGISTER_ ACK message from the ONU/ONT, the ONU/ONT is registered in the EPON system and the discovery process is complete.

The next MAC task is the negotiation of bandwidth between OLT and ONU/ONT. Without specifying any particular bandwidth allocation algorithm, MPCP provides the REPORT/GATE mechanism to manage the EPON upstream bandwidth. In order to assign upstream bandwidth dynamically, the OLT monitors the incoming traffic of each ONU/ONT and makes an arbitration decision in accordance with the traffic load and other QoS requirements. The REPORT and GATE MAC control messages are used to facilitate bandwidth negotiation.

As exemplified in Figure 3.27, an ONU/ONT sends a REPORT message to the OLT containing its local time stamp (TS) and queue status. The OLT calculates the round-trip time (RTT) from the reported time stamp (TS). In contrast to BPON and GPON, it is the ONU that measures the RTT and thus also has precise timekeeping. Upon making the allocation decision, the OLT sends a GATE message downstream containing the information of the TS, grant start time, and grant length. As allocations vary; the GATE message stream may or may not be a relatively steady flow, depending on the bandwidth allocation/assignment algorithm. The destined ONU updates its local clock by the received TS and transmits its data from the grant start time in the grant length. Different manufacturers are free to implement different DBA algorithms to manipulate the ONU/ONT queue status report and QoS metrics.

Energy conservation, the heart of "green" communications, has become a significant element in the design of MAC layer control mechanisms. The large increase in capacity between 1-Gbps EPON and 10-Gbps EPON carries with it a substantial, if not quite proportional, increase in energy consumption. In addition to higher transmitter power, there are more functions to power, such as the mandatory FEC and faster processing of the received signal. To reduce the energy cost, researchers have been exploring "sleep" modes for these high-capacity networks in which network equipment only consumes significant energy when it is actually needed for high-speed communication [ANSARI]. To maintain service quality, MAC layer control and scheduling schemes must be designed to disable and enable functions quickly.

As noted in Chapter 1, the largest potential energy savings are in the ONUs, so that is where sleep–wake functions are implemented. An idle ONU cannot be totally asleep because it must be able to receive and act on notification that traffic is about to be sent to it. Providing this "asleep but aware" capability is not the greatest challenge; it is in the complexity and latency of that part of the MPCP between OLT and ONU for waking up the ONU, all the while queueing the downstream traffic arriving from the network for that ONU. Compliance with the MPCP standard requires that sleeping ONUs wake up every 50 ms to send an MPCP REPORT message and receive a GATE (TS and grant of time slots) message.

As suggested in [ANSARI], it appears that an effective way to implement a sleep mode without degrading performance is to put different elements of an ONU to sleep in different situations rather than putting the whole ONU to sleep. Specifically, if there is no traffic in either direction, every component in the ONU can sleep, while if there is only downstream traffic, only those functions related to upstream transmission sleep, and similarly, if there is only upstream traffic, then the downstream functions can sleep. Periods of upstream traffic may actually be intermittent, with only some time slots used, so that there the laser functions can sleep some of the time. Over all operational modes, the result is a dynamically varying power utilization, ideally minimizing the average power.

3.4.4 Comparison of 1G EPON and GPON

Table 3.6 summarizes the major features of GPON and 1G EPON, and a similar comparison could be made for XG-PON and 10-Gbps EPON. EPON is essentially an extension of Ethernet technology from LANs to the access network domain. Its features include simple implementation and compatibility with existing low-cost Ethernet equipment. EPON is a promising solution for densely populated areas and the overlaid network market and is already firmly entrenched in Asia. By the end of 2008, EPON accounted for 60% of all FTTx subscribers worldwide, while GPON accounted for only 17% [IDATE]. In the Asia Pacific region, EPON reached 91% of the FTTx subscribers, while the remaining 9% used point-to-point Ethernet. In North America, on the other hand, GPON accounted for 74% of FTTx subscribers and EPON for only 5%.

As the evolved version of BPON, GPON inherits the attractive features of broadband access from BPON. It has protection switching, encryption, and strong error correction in its CRC codes, in comparison with EPON's lack of the first two and relatively weak error correction capability. By introducing the GEM mechanism, GPON can provide more flexible services with improved efficiency. Its enhanced transmission rate and compatibility with ATM technology make GPON the leading PON technology in North America. EPON may have advantages in its use of a large address space (a component can keep its address despite future changes in network configuration) and in its freedom from 125-μs framing with its associated complexity for traffic integration [Green] (Table 3.7).

Both GPON and EPON are viable broadband access systems, with deployment depending on the required rates and QoS, and the expected costs. A cost comparison [Parsons] made in 2005 claims that GPON enjoys a significant cost advantage (Figure 3.28) as a result of GPONs "higher split ratio, PON bandwidth and bandwidth efficiency." However, the ubiquity, familiarity, and low cost of Ethernet equipment make it attractive for mass deployment and may imply cost advantages not evident from a focus on the use of fewer OLTs.

TABLE 3.7   Features of EPON and GPON ([Green], [Parsons])

Features	EPON	G-PON
Standard	IEEE 802.3ah-2004	ITU-T G.984.x
Line rate	1.25 Gbps symmetric	Max 2.448 Gbps symmetric
Revenue BW*	900 Mbps	2300 Mbps
Downstream efficiency*	72%	92%
Upstream wavelength	1310 nm	1310 nm
Downstream wavelength	1490 nm	1490 and 1550 nm (optional)
Data unit	Ethernet frame	GEM frame, ATM cell
Split ratio	1:16, 1:32 (with FEC)	1:64 (or possibly 1:132)
Max distance	20 km	20 km
TDM support	Ethernet frame encapsulation	Directly
QoS support	802.1Q priority levels	Fixed, assured, nonassured, best effort
Address space	48 bits	8 bits
Class of service	8 queues	5 T-CONT types
Security	Not specified	AES encryption
Protection	Not specified	Four specified architectures

* The actual bandwidth of data transmission.

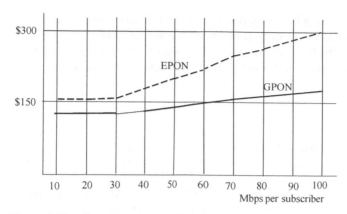

Figure 3.28 Cost comparison of GPON and EPON [Parsons].

3.4.5 Service Interoperability in EPON (SIEPON)

Network operators who want a choice among vendors of network equipment including the use of smoothly interworking equipment from multiple vendors, welcome interoperability initiatives in standards. One promising initiative begun in 2011 is SIEPON, designed "to build upon the IEEE 802.3ah (1G-EPON) and IEEE 802.3av (10G-EPON) Physical layer and Data Link layer standards and create a system-level and network-level standard, thus allowing full 'plug-and-play' interoperability of the transport, service, and control planes in a multivendor environment" (http://standards.ieee.org/develop/wg/SIEPON.html). This IEEE P1904.1 standard-in-progress complements the existing IEEE 802.1 and 802.3, which address interoperability at the physical and data link layers (http://grouper.ieee.org/groups/1904/1/).

The P1904.1 Working Group was, at the time of writing, planning work on "system-level interoperability specifications covering equipment functionality, traffic engineering, and service-level QoS/CoS mechanisms," together with "management specifications covering equipment management, service management, and power utilization" (http://standards.ieee.org/develop/project/1904.1.html). As noted by at least one analyst, "SIEPON is an unusual standards-working group for the IEEE. Rather than focusing on technical solutions, it focuses on the sharing and reutilization of best practices in networks supporting more than 40 million EPON-based FTTx subscribers. SIEPON participants include carriers . . . and chip vendors . . . It is unusual for major carriers . . . to share their specifications, test protocols, and findings [but] this sharing will, enable them to influence the vendors' offerings, in essence creating a worldwide standard for service interoperability with conformance testing and a planned certification program. . . . An international standard and compliance program should lead to lower cost equipment for all." [OVUM]

Although EPON is not necessarily going to overcome the considerable sales and deployment lead of GPON in some markets, this attention to interoperability will surely have an impact. It also suggests a growing trend in standardization generally, joining traditional technologies at the lower protocol layers with higher-layer considerations to further strengthen the union of information and communication technologies in the modern world.

REFERENCES

[ANSARI] J. Zhang & N. Ansari, "Toward energy-efficient 1G-EPON and 10G-EPON with sleep-aware MAC control and scheduling," IEEE Commun. Magazine, February, 2011.

[Capurso] A. Capurso, R. Mercinelli, M. Valentini, & M. Valvo, "Method for transmitting data packets with diffrerent precedence through a passive optical network," WIPO pub. WO/2007/051488, May 5, 2007.

[Green] P. Green Jr., *Fiber to the Home: The New Empowerment*, Wiley, 2006.

[IDATE] "FTTx Market Report," IDATE Consulting & Research, July, 2009, available at http://www.telecomasia.net/pdf/ZTE/ZTE_093009.pdf

[McDysan] D. McDysan & D. Spohn, *ATM Theory and Applications*, McGraw-Hill, 1998.

[OLA] T. Orphanoudakis, H.-C. Leligou, & J. Angelopoulos, "Next generation Ethernet access networks: GPON vs. EPON," 7th WSEAS Internat. Conf. on Electronics, Hardware, Wireless & Optical Communications, Cambridge, UK, February, 2008.

[OVUM] "SIEPON a big step forward in FTTx standardization," March 7, 2011, available at http://ovum.com/2011/03/07/siepon-a-big-step-forward-in-fttx-standardization/

[Parsons] D. Parsons, "GPON vs. EPON costs comparison," Lightwave, September, 2005.

[RFC4836] E. Beili, "Definitions of Managed Objects for IEEE 802.3 Medium Attachment Units," April 2007.

[Sangha] J. Sangha, "Technology overview broadband passive optical networks (BPON)," NCTIA Conference, June, 2005.

[TANAKA] K. Tanaka, A. Agata, and Y. Horiuchi, "IEEE 802.3av 10G-EPON Standardization and Its Research and Development Status," IEEE/OSA J. Lightwave Tech., 28(4), pp. 651–661, February 15, 2010.

4

RECENT ADVANCES AND LOOKING TO THE FUTURE

Passive optical networks (PONs) are accepted as a high-performance, modest-cost choice for high-speed broadband access as the public network extends optical transmission to the network edge. The progress being made now and in the coming years will result in enhanced interoperability among equipment from difference manufacturers; expanded capacity through wavelength division multiplexing (WDM); longer reach; new interfaces with evolving carrier networks, particularly with Carrier Ethernet; and imaginative new architectures such as wireless/optical integration. This final chapter describes what these initiatives, in both the physical and higher protocol layers, will mean for future broadband communications.

4.1 INTEROPERABILITY

Although an operator will generally work with a single vendor, PON system interoperability encourages service consistency across operators and the flexibility available to operators for future purchases and enhancements. As with many other technologies, interoperability encourages mass deployment by multiple operators and customer acceptance.

Without interoperability assessment to resolve common issues within the PON industry, PON equipment vendors will tend to implement diverse

The ComSoc Guide to Passive Optical Networks: Enhancing the Last Mile Access,
First Edition. Stephen Weinstein, Yuanqiu Luo, Ting Wang.
© 2012 Institute of Electrical and Electronics Engineers. Published 2012 by John Wiley & Sons, Inc.

interpretations of PON standards and recommendations. Interoperability assessment enables a more uniform implementation of various PON products, facilitating interoperation across multiple vendors. Operators buying the equipment can then efficiently manage a diverse group of optical network units (ONUs)/optical network terminals (ONTs) and optical line terminals (OLTs), and competition among vendors on similar products encourages a high quality-to-price ratio. We can summarize the benefits of interoperability as the following:

Accelerating global acceptance of PON technologies

Fostering industry-wide competition and lower-cost PON products

Reduced implementation cost

A larger selection of PON solutions for service provisioning

Building the long-term viability of the PON platform

Full Service Access Network's (FSAN) Interoperability Task Group, together with the International Telecommunication Union (ITU), hosted a series of broadband passive optical network (BPON) and gigabit-capable passive optical network (GPON) interoperability events over the years, evaluating both 1:1 and 1:N testing (see the next subsection). The first interoperability tests, on BPON, were held in Tokyo, Geneva, and San Ramon (California) in 2004 [Lightwave2004]. The tests in Tokyo and Geneva "demonstrated BPON support for broadband data services over multivendor BPON networks from six vendors" and also verified conformance at the transport layer to the International Telecommunication Union-Telecommunication (ITU-T) G.983 recommendations. The San Ramon demonstration, at the SBC Laboratories, "involved voice testing between four brands of OLT central office equipment . . . and eight brands of ONT subscriber equipment," transporting voice signals between telephone interfaces on the ONTs and switch interfaces on the OLTs. These may not have been full-feature tests but did demonstrate interoperability for basic services.

Leveraging the lessons learned from BPON interoperability, FSAN and the ITU launched GPON interoperability tests at Telcordia Technologies in January 2006. Five system and device vendors interconnected their GPON products, identifying and resolving GPON transmission convergence (GTC) layer incompatibilities. The second test in May involved 10 participating GPON vendors, and a third GPON interoperability test was held by KTL, a commercial test laboratory (http://www.KTL.com) in California, in September. At this event, 11 PON vendors evaluated the interoperability of the ONU/ONT management and control interface (OMCI) through a set of detailed test cases. In early December, there was a public GPON interoperability demonstration at the ITU Telecom World 2006 exhibition in Hong Kong [Telcordia GPON 2006], with 10 participating GPON vendors and 5 supporting operators from around the world. Voice, video, and data services were tested, as indicated

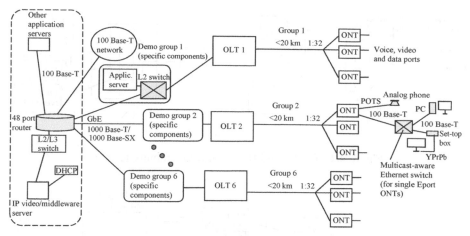

Figure 4.1 GPON interoperability configurations tested at ITU Telecom World 2006. [Telcordia GPON 2006]. POTS, plain old telephone service; DHCP, dynamic host configuration protocol.

in Figure 4.1, and similar tests were held in later years [Telcordia GPON 2008]. ITU-T GPON interoperability demonstrations, increasingly specialized, have become frequent events, including ITU's October 2010 Plugtest "to verify behaviour of OLTs and ONUs in error recovery cases and fault situations" [ITU Plugtest 2010]. These tests aim at both resolving problems and testing new capabilities.

EPON has had sporadic interoperability tests. The first known to the authors was at Supercomm 2004, simultaneous with ratification of the Institute of Electrical and Electronics Engineers (IEEE) 802.3ah Ethernet in the First Mile Standard [TMCnet EFM 2004]. Vendors have hosted later interoperability demonstrations, such as the demonstrations of interoperability of symmetric 10G-EPON products held in Shanghai, China in April 2010 [10GEPONINTEROP].

Interoperability issues can be difficult to resolve. Section 4.1.2 discusses some of the hurdles and paths to solutions.

4.1.1 Implementing 1:1 and 1:N Interoperability Testing

One-to-one (1:1) interoperability testing interconnects ONUs/ONTs from only one vendor with the OLT from another. It analyzes how well the PON products interwork between different vendors without special effort on the part of PON subscribers. Figure 4.2a illustrates the setup. The 1:N interoperability test setup shown in Figure 4.2b is more complicated, with all the ONUs/ONTs and the OLT from different PON vendors.

Both 1:1 and 1:N testing architectures evaluate functions supported by multiple PON layers. In GPON, these layers include the physical media dependent (PMD) sublayer, the transmission convergence (TC) sublayer, the

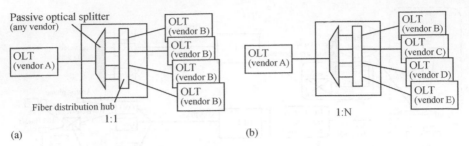

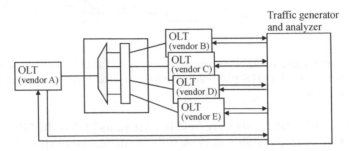

Figure 4.2   (a) 1:1 and (b) 1:N interoperability testing.

Figure 4.3   Test setup with traffic generation and analysis.

ONT management and control interface (OMCI) sublayer, and the application layer.

The test cases for PON interoperability should address the areas of high interest to PON operators such as optical compatibility (physical layer requirements such as wavelength accuracy, signal strength, and synchronization), ONU/ONT turnup and management, and service-related functionality. In particular, the following two important services need to be tested in an interoperated environment:

- voice over Internet protocol (VoIP), including the two well-defined standards implementations: the ITU-T Recommendation H.248 and the Internet Engineering Task Force's (IETF) Session Initiation Protocol (SIP)
- Internet protocol television (IPTV) over multicast, with the IETF Internet Group Management Protocol (IGMP) implemented in the PON products, controlling video multicast flows through the interconnected PON system

A general test setup is illustrated in Figure 4.3. The traffic generator/analyzer is used to input voice, data, and video traffic into the PON system and to detect the received traffic at the two ends. The monitored parameters associated with each service include packet loss, delay, and network throughput.

4.1.2   Management and Quality-of-Service (QoS) Challenges

PON management poses the most difficult challenge for PON interoperability. In BPON and GPON, the concern is with the OMCI (ONT management and control interface), while in EPON, it is the MPCP (multipoint control protocol). PONs transfer management information via a set of control messages, which are either embedded in BPON or GPON upstream/downstream frames or are carried by Ethernet frames. In either case, the interoperability test tools must support an "on-PON" analysis functionality that can extract control messages from the traffic stream to test the provisioning of dynamic bandwidth allocation (DBA). This functionality, necessary to validate the proper operation of the vendor-defined resource allocation algorithm, has not always been available in commercial traffic analysis tools. As it becomes more generally available, additional test cases will be required to more fully evaluate PON interoperability.

QoS, which is interpreted differently in different PON products, represents a second challenge for interoperability testing. Different queue management schemes may be implemented at the ONU/ONT, following the different preferences of PON vendors. At the OLT, diverse scheduling schemes may be found among ONUs/ONTs from different vendors. These differences make it difficult to test, and to provision in actual operator networks, consistent service quality and service differentiation guarantees.

4.2   WAVELENGTH DIVISION MULTIPLEXED PON (WDM-PON)

By 2020, the broadband access requirement is very likely to be at least 1 Gbps of continuous bandwidth per residential user. This will support services such as ultra-HDTV [Kuroda] with a resolution up to 7680×4320 pixels and 24-channel audio. A very high-speed WDM-PON appears to be the most promising solution, in terms of both cost and performance, for this large per-subscriber requirement. A detailed analysis [Grobe] suggests that ultra-densely spaced WDM, with wavelengths at less than 12.5-GHz spacing, and a simple WDM-PON architecture using filters rather than splitters/combiners, will be required to minimize both cost and energy consumption. This structure will serve 128–192 clients over distances up to 60 km.

The attractions of using WDM include a large increase in capacity and the simplicity and privacy associated with dedicating a particular wavelength to a particular customer, although most customers will require less than the entire capacity of a wavelength channel [Banerjee]. Customers can build virtual private networks using reserved capacity such as dedicated wavelengths or smaller-granularity allocations provided by subcarrier or time division techniques. WDM-PON, shown in Figure 4.4, which comes in different wavelength densities described in the next subsection, appears to be the next-generation wired broadband access solution.

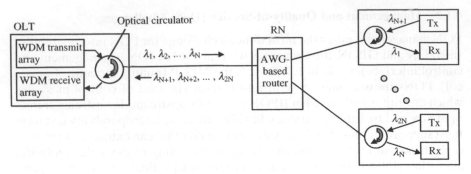

Figure 4.4 WDM-PON.

There are, fortunately, several approaches to making finer allocations of capacity. These include the conventional TDM/time division multiple access (TDMA) techniques. Another is subcarrier multiplexing (subcarrier modulation [SCM]-WDM), including multiple access orthogonal frequency division multiplexing (OFDM-WDM), which allows users at different ONUs to send upstream signals in different frequency slots around the same wavelength within the WDM set. A third approach is to use optical code division multiplexing (OCDM-WDM), which allows sharing the transmission resource of a particular wavelength among multiple unsynchronized users without using a central controller. Each user invokes a unique code, in the spirit of a classical time sequence spread spectrum that uniquely identifies the user and conveys user data upstream. [Beyranvan].

Adding an additional dimension of capacity to the TDM-PONs discussed up to now, which use TDM downstream and TDMA upstream for medium access, WDM-PON makes possible wavelength division multiple access (WDMA). Each ONU/ONT can be assigned a unique wavelength pair for communication with the OLT, one wavelength for downstream and the other for upstream, providing a separate point-to-point connection between each ONU/ONT and the OLT through the shared point-to-multipoint physical architecture. The term "colorless ONU" is sometimes used for an ONU in which the upstream wavelength is not fixed but rather tunable as required [Lam]. This permits wide use of the same physical unit in different systems and configurations.

ONUs/ONTs are capable of communicating independently with the OLT in different rates and data formats. Privacy of the communication link is easier to provide since ONUs/ONTs in WDM-PON receive only on their own assigned wavelength and do not receive the data sent on different wavelengths to other users. This privacy is still dependent on a high degree of optical isolation among the ONUs and the wavelength channels.

Of course, it is possible to combine WDMA and TDM/TDMA, or WDMA and subcarrier channels (Section 4.3), to realize finer granularities, with a wavelength pair shared among several users.

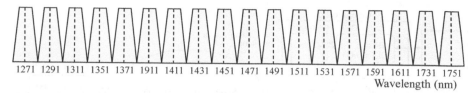

1271	1291	1311	1351	1371	1911	1411	1431	1451	1471	1491	1511	1531	1571	1591	1611	1731	1751

Wavelength (nm)

Figure 4.5 Coarse wavelength division multiplexing (CWDM) wavelength assignments.

4.2.1 Coarse Wavelength Division Multiplexing (CWDM)-PON and Dense Wavelength Division Multiplexing (DWDM)-PON

With respect to wavelength assignment, WDM-PON can be categorized as CWDM-POM and DWDM-PON. CWDM-PON follows ITU-T Recommendations G.695 and G.694.2 (available at http://www.itu.int/rec/T-REC-G/e). It uses the spectrum of 1271–1611 nm with 20-nm wavelength spacing and provides 18 wavelength channels, as shown in Figure 4.5. Because of the coarse granularity, strict tuning of wavelength is not required for data transmission through CWDM-PON, and systems can employ low-cost arrayed waveguide gratings (AWGs) without thermal regulation and uncooled lasers. The selection of a particular type of laser depends on the system requirements (e.g., rate per ONU and wavelength plan), but the more expensive distributed feedback (DFB) lasers are more likely to be used in the downstream direction, while Fabry–Perot (FP) lasers may be more practical for upstream transmission.

DWDM-PON adopts a finer granularity, with the typical wavelength spacing less than 3.2 nm. An option is to use the DWDM spectrum specified by ITU-U Recommendation G.692 (http://www.itu.int/rec/T-REC-G/e), which ranges from 1528.77 to 1563.86 nm with a center wavelength of 1553.52 nm. In comparison with CWDM-PON, DWDM-PON features longer reach, 70–135 km in experimental systems [LeeMunLee] [Davey], and more supported ONUs/ONTs. On the other hand, DWDM-PON imposes higher costs for crosstalk avoidance between the narrower wavelength channels.

The increasing demand for access bandwidth stimulated increased industrial/university collaborative activity to overcome the cost and performance obstacles of DWDM-PON. Europe initiated the innovative Scalable Advanced Ring-Based Passive Dense Access Network Architecture (SARDANA) project [SARDANA1] [SARDANA2] [Teixeira] among major government, industry, and educational institutions to advance the state of the art and to contribute to international fiber-to-the-home (FTTH) standards. There was an early test-bed demonstration in Espoo, Finland, including extended scalable reach; a field trial in Lannion, France, in 2010 that demonstrated delivery of broadband services over a large metropolitan area; and a public demonstration at the FTTH European Union (EU) Council meeting in Milan in February 2011.

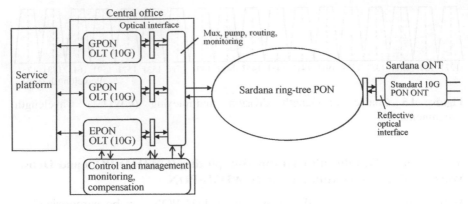

Figure 4.6 High-level SARDANA system architecture [SARDANA2].

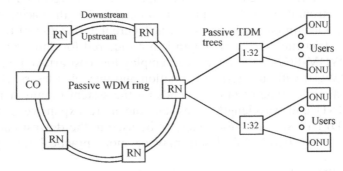

Figure 4.7 SARDANA hybrid passive ring-tree architecture [Texeira].

Figure 4.6 illustrates the SARDANA 32-wavelength WDM/TDM fully passive fiber ring PON (expanded in Figure 4.7) supporting communication between multiple OLTs, both GPON and EPON, and multiple user locations. The experimental system as of late 2010 had 16 RNs, each of which supported two trees, each serving 32 users, implying 1024 user endpoints. The ONTs operate at 10 Gbps downstream and 2.5 Gbps upstream on each wavelength, typical of a system based on GPON, but SARDANA is transparent to different transmission technologies. The objective is a very high-capacity system that maximizes the number of users, the size of the served area (greater metropolitan out to 100 km), and the capacity available to each user location, 100–1000 Mbps downstream for each user, while minimizing infrastructure and component cost. It is compatible with both the GPON and the EPON medium access control (MAC). It has a multioperability model supporting operators of different types of PON at protocol layer 2 and a variety of service operators at level 3.

The SARDANA architecture is a hybrid combining WDM ring and TDM access tree topologies joined at a specially designed remote node (RN), as

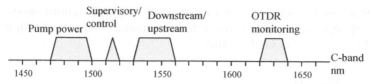

Figure 4.8 SARDANA wavelength allocation [SARDANA2]. OTDR, optical time domain reflectometer.

shown in Figure 4.7. New RNs are easily added, supporting scalability. The ring employs two fibers, one for downstream traffic and the other for upstream traffic, "in order to optimize the spectrum and avoid Rayleigh backscattering distortions" [Texeira]. Both the ring and the tree are passive, making their combination a new kind of hybrid PON. Each of the two trees on an RN is served by its own wavelengths so that each RN delivers two downstream wavelengths and picks up two upstream wavelengths (which are the same wavelengths as described below). Figure 4.8 shows wavelength allocation. A single fiber is employed in each TDM tree.

Although the ring is passive, with all optical signal generation at the central office (CO), there is signal amplification at each RN, needed to overcome attenuations from transmission, add/drop, and filtering elements of the system. An erbium-doped fiber amplifier (EDFA) is remotely pumped from the CO through the upstream fiber, and the system can also employ Raman amplifiers. The overall capacity (number of RNs) of the network is limited by the total pump power.

The user location is similarly passive. The ONU employs a wavelength-independent reflective semiconductor optical amplifier (RSOA) that generates the upstream carrier by reusing a downstream signal. The wavelength of the upstream signal is thus identical to that of the downstream signal. Directional couplers minimize upstream/downstream interference.

The SARDANA architecture illustrates an extension of the traditional concept of a PON that can potentially realize huge access capacities. Initial evaluations [SARDANA1] indicate capital expense (CAPEX) that is considerably lower than GPON at all population densities. Inexpensive, mass-produced optical devices are a necessity if this experimental design is to become a standard architecture for the residential market.

4.2.2 WDM Devices

At the OLT, WDM-POM uses a WDM source to generate a set of wavelengths, ordinarily one for each ONU/ONT, although a single wavelength could serve multiple customers, particularly with subcarrier multiplexing that is generated in the electrical modulation signal (Section 4.3). At each ONU/ONT, an optical filter selects the particular wavelength carrying downstream data to that user.

At the RN, optical components with multiplexing and demultiplexing functions enable upstream and downstream transmission, respectively. Options for the WDM source in an OLT include

Multifrequency laser
Gain-coupled DFB laser array
Chirped-pulse WDM source

Options for the colorless wavelength transmitter in an ONU/ONT include

Tunable laser
Vertical-cavity surface-emitting laser diode
Injection-locked laser
Reflective-type semiconductor optical amplifier (RSOA)

Section 2.2.3 described the basic properties of an AWG. For the RN, an athermal (without temperature control) AWG, an example of which is shown in Figure 4.8, appears to be the most promising component, providing low-cost wavelength multiplexing and demultiplexing when, as in WDM-PON, extreme precision is not a requirement. The cyclic wavelength property of an AWG enables it to function simultaneously as a multiplexer in one direction and a demultiplexer in the opposite direction. Figure 4.9 shows other advanced optical components as well. The Infinera 100-Gbps photonic integrated circuit (PIC) (http://www.infinera.com/technology/pic/largescale.html) produced 10 wavelengths each conveying 10 Gbps.

A WDM-PON system can be realized with the same relatively inexpensive injection-locked laser signal generator used in both OLT and ONT/ONU.

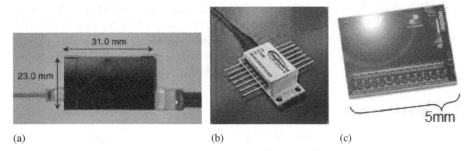

(a)   (b)   (c)

Figure 4.9 Advanced optical devices. (a) NTT Photonics' athermal AWG [Kamei], (b) Eagleyard DFB laser diode (www.eagleyard.com/en/products/dfb-dbr-laser), (c) Infinera 100Gbps photonic integrated circuit transmitter (www.infinera.com/pdfs/whitepapers/Photonic_Integrated_Circuits.pdf). There are also later 500Gbps PICs.

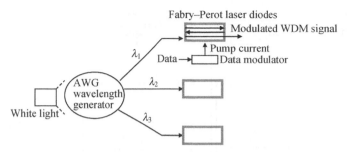

Figure 4.10   Identical Fabry–Perot laser diodes generating WDM modulated signals.

Instead of wavelength-specific signal generators for the different wavelengths, the laser sources are identical [Fabry–Perot] laser diodes (FP-LDs), each wavelength-locked to the desired wavelength. One wavelength locking, illustrated in Figure 4.10, works as follows [LeeMun]: The data-modulated pump current modulates an amplified carrier waveform bouncing within the cavity of an FP-LD. The carrier wavelength is controlled by that of an injected optical signal, which is one of a set of typically 16 or 32 wavelength-selective signals generated by applying an unmodulated broadband light source to an AWG wavelength splitter. The relatively strong carrier, generated through optical injection "seeding" by a relatively weak driving signal, exploits the semiconductor gain of the FP-LD. Injection locking is a relatively low-cost technique used primarily for modest bit rates and transmission distances, but careful design is needed to wipe out the downstream modulation on the injected carrier [Cheng].

FP lasers have shown good performance, especially with regard to thermal reliability, in WDM-PONs. One study reported error-free transmission at 155 Mbps extended over a temperature range of 70°C as may be found in ONTs [Shin].

High cost remains the main obstacle in the implementation of WDM-PON, which will advance with the realization of low-cost devices for WDM sources, receivers, and wavelength routers. There are also technical challenges including improved performance for RSOAs and FP lasers (not as good as the more expensive tunable laser), and optimization of RSOA gain for different fiber drop lengths [Cheng]. Korea Telecom has led the deployment of WDM-PON, beginning with field trials in 2004 and volume deployment in 2005. WDM-PON holds great promise as the next-generation passive optical network (NG-PON) technology for high capacity and relatively low cost per subscriber.

4.3   SUBCARRIER PON

The granularity of physical channels in a PON can be considerably finer than that of WDM. One increasingly useful approach is SCM-PON, in which a

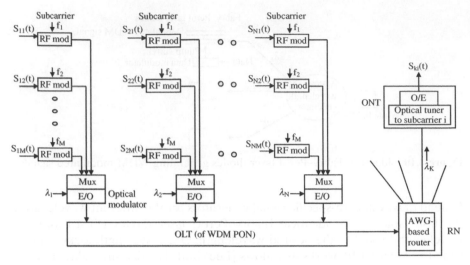

Figure 4.11 Downstream subcarrier multiplexing in a WDM-PON.

number of frequency division multiplexed subchannels are modulated onto a single carrier, as suggested in Figure 4.11, for an SCM/WDM-PON (only downstream shown). At the customer's location, an individual SCM signal, shown as $s_{ki}(t)$, representing the ith modulation on the kth wavelength, can be recovered with a narrowband optical tuner.

The SCM spectrum may be either double sideband (from simple amplitude modulation of the optical carrier) or single sideband as in OFDM (downstream) or OFDMA (upstream) systems [Qian]. OFDM/OFDMA signals are generated electrically using the fast Fourier transform (FFT) computational technique and then modulated on optical carriers.

The concept of SCM of optical carriers has been known for a long time, with relatively early application in hybrid fiber-coax cable television (CATV) systems [Darcie]. It is not even necessary that the modulated subcarrier signals be first electrically combined in a Mux as shown in Figure 4.11 before E/O conversion (optical modulation). Optical modulation may be performed before multiplexing, with the optical modulators either colocated in the OLT or implemented elsewhere, or geographically dispersed, perhaps cascaded as demonstrated in Figure 4.12 [Shibutani_et al.]. The optical modulation index (OMI), which is defined as the peak-to-peak modulation amplitude divided by the average light power, must be kept below 4% or 5% per subcarrier to adequately suppress second- and third-order mixing terms in the frequency spectrum [Salim].

SCM is suitable for support of "triple play" services. An effective architecture, shown in Figure 4.13, optically modulates a video signal on an optical subcarrier, uses an optical interleaver to separate this modulated subcarrier from the carrier itself, modulates the carrier with downstream data, and

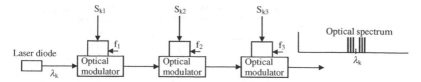

Figure 4.12 Cascaded optical modulators for an SCM transmission system with a two-sided spectrum.

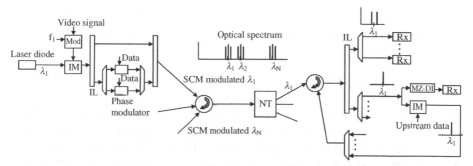

Figure 4.13 Use of subcarrier modulation and centralized lightwaves for economical triple-play service through an SCM/WDM-PON [Yu1]. IL, insertion loss; IM, intensity modulator.

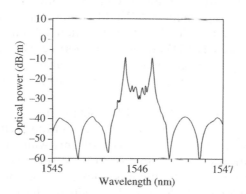

Figure 4.14 Signal spectrum transmitted in the experimental apparatus of Figure 4.12 [Yu1].

combines them in an optical interleaver for downstream transmission along with other subcarrier-modulated optical wavelengths [Yu1]. The data-modulated carrier is cleaned up in the ONT in order to be used by the ONT for upstream data transmission, an example of the "centralized lightwave" technique that can substantially reduce ONT costs by eliminating a laser diode. Figure 4.14 shows the spectrum (two-sided in this case), carrier and subcarriers, around one optical wavelength, 1546 nm.

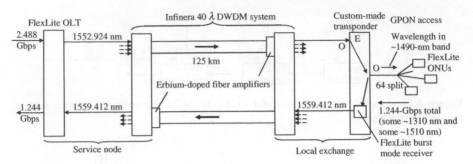

Figure 4.15 An experimental 135-km DWDM GPON configuration.

4.4 LONG-REACH PON

The objective of long-reach PON is "to use a high capacity, high split PON, with a reach in excess of ~100 km, to combine optical access and metro (metropolitan networking) into a single system" [Davey]. This makes PON more than just an access networking option. This reach requires a transponder with optical amplifiers between the OLT and the RN, as indicated in Figure 4.15. The implementation of this long-range system may benefit from "a high degree of optical component integration made possible by the use of large-scale, monolithically integrated PICs to consolidate the key optical components required to implement a DWDM component, effectively a WDM system on a chip." The long haul part of the extended-reach PON is actually a standard high-capacity DWDM optical transport network, isolated from the end user, benefitting from economies of scale realized for long-haul optical networks, and there is no exposure of end users to possibly dangerous laser radiation.

The experimental system of Figure 4.15, using standard G.652 fiber, realized a round-trip delay of 2.25 ms for an E1 (2.048 Mbps) client, meeting the GPON standard, so that there were no problems with the delay or timing recovery. This illustrates the potential of PON to serve large populations of users far from a common OLT. Of course, newer, more powerful PON architectures such as SARDANA (Section 4.2.1) already have a remarkably long reach.

4.5 OPTICAL–WIRELESS INTEGRATION

As noted in Chapter 1 and in books such as [Green], the access network is the bottleneck for broadband service provisioning. The proposed high-speed solutions to this problem exist in three categories: optical access technologies such as PONs; copper access techniques such as coaxial cable (http://www.cablelabs.com), short-reach digital subscriber line (DSL), and multiuser systems based on twisted-pair subscriber lines [Cioffi et al.]; and wireless technologies such as Worldwide Interoperability for Microwave Access (WiMAX) (http://www.wimaxforum.org).

There are many reasons why a *combination* of optical and wireless technologies can be advantageous. These reasons include

- backhaul of data traffic from wireless base stations (BSs) or access points (APs) to concentration points or switching centers;
- "radio over fiber" (RoF) transport of wireless signals from BS/APs to centralized processing points, reducing the complexity of the many BS/ AP units to simple power amplifiers; this realizes a synergy of optical fiber capacity and wireless communications mobility with the possibility of reduced costs;
- wireless/optical mesh networks, for example, a metropolitan network composed of a wireless mesh network with many interfaces to the metropolitan optical backbone, providing an attractive combination of mobility, capacity, and flexibility;
- mutually protective overlays, in which, for example, overlaid PON and WiMAX access systems provide emergency capacity in the event of failures; and
- application of wireless techniques and modulation formats, such as OFDM, [Hanzo et al.] to realize advantages against chromatic and polarization mode dispersons [DXW] or greater flexibility such as finer granularity in access systems.

With these applications in mind, a number of system architectures and technology choices are under consideration. There is a growing belief that optical–wireless integration helps enable the extension of the high-bandwidth backbone network to the distribution network and to subscribers. The following subsections show how it is done.

4.5.1 Architecture

Figure 4.16 illustrates backhaul and overlay possibilities. It employs PON technology as one phase of an access network. The OLT, possibly but not necessarily colocated with the optical edge node, enables upstream local data aggregation and downstream data broadcast. Each ONU connected to the PON's RN can support wired access, wireless access, or both.

The integrated architecture in Figure 4.16 connects subscribers to the core network edge node in two steps. The first step uses the existing PON connections to the curb and neighborhood, while the second step extends services to the subscribers by employing either optical fiber or wireless WiMAX links. In the figure, (A) illustrates an ONU supporting both normal PON access for a residence and WiMAX access for a business. A WiMAX BS is colocated with

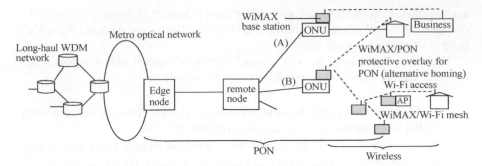

Figure 4.16   Several forms of optical–wireless integration.

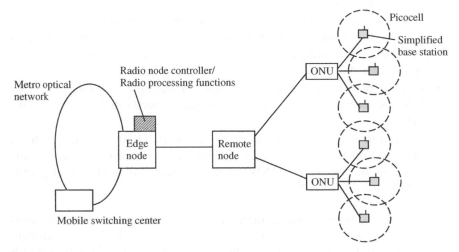

Figure 4.17   PON supporting RF over fiber for simplified cellular mobile picocell base stations and aggregated RF processing.

the ONU. Access link (B) supports only a WiMAX BS, shown here providing network access to a WiMAX/Wi-Fi mesh network and also providing WiMAX protection backup to the residence normally using ONU (A). If that ONU should fail, the residence uses its WiMAX link to ONU (B) as an alternate homing structure.

Figure 4.17 illustrates support of wireless (typically cellular mobile) picocells through carriage of radio frequency (RF) signals between an optical edge node enhanced with wireless processing functions and the picocell BSs [Pinter1] [Yu2]. All signal processing functions including modulation, detection, coding, and RF channel equalization can be concentrated in the enhanced optical edge node, offering economy of scale and considerable simplification of the BSs.

RoF frequently carries the signals for several BSs over a fiber, as is the case in Figure 4.17. Techniques such as OFDM can efficiently generate the subchannels for these signals in the overlay scenarios described in the next subsection. These signals are, however, subject to distortions in the optical path (largely intermodulation distortion of the E/O and O/E conversions arising in part from nonlinearities in lasers and amplifiers) as well as the wireless link. Equalization of the combined wireless/optical transmission path may be required, for example, in the combination of a polynomial filter and decision feedback equalizer described in [Pinter2].

4.5.2 Integration Modes, Benefits, and Challenges

Wireless and PON share the common challenge of resource allocation. Once the physical level is established, the system designer must consider how to map resource allocations in the wireless domain, such as noncontention transmission time slots and selective retransmission backoff in contention intervals, into resource allocations in the optical networking domain, such as wavelengths, subcarriers, and dynamic bandwidth assignment. An example of the MAC-level interoperability challenges is provided later in this section.

The integration of the transport of wireless signals with other PON services can be addressed in several ways. One approach, shown in Figure 4.18a, is to directly connect the wireless infrastructure to a PON termination (NT). The wireless traffic is treated as a portion of the PON digital traffic and served as one among several possible PON clients. In this arrangement, the radio signals must be digitized, which, in the absence of compression, may require

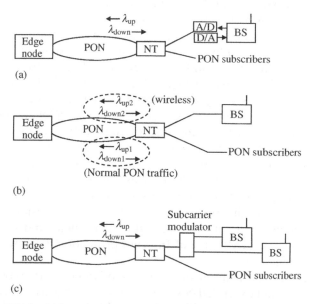

Figure 4.18 PON–wireless integration modes. (a) In-PON digitized; (b) analog overlay; (c) analog on subcarrier. NT, network termination.

transmission data rates considerably larger than the rates of the information signals imbedded in the radio signals. An intermediate frequency (IF) rather than RF modulation may be used, which is reconstructed into an RF signal at the O/E conversion point after optical transmission. The digitized radio signals must conform to the PON management mechanisms, and data encapsulation is necessary to map the wireless data into the PON domain. This option is called "in-PON integration."

An alternative mode, possibly exploiting the extra wavelength provided for video services in the GPON and EPON standards, is to overlay the wireless signal through the PON connection as shown in Figure 4.18b. In this option, one more wavelength is assigned to carry the wireless signal, opening a new transmission pipe in the original PON without consuming the normal PON bandwidth. The parameters of the wireless traffic can be entirely different from those of the normal PON traffic, and in particular, the wireless signals can be left analog as suggested in the figure.

An additional overlay option, shown in Figure 4.18c, is the use of SCM in a single wavelength system. Here, the wireless signal is modulated into the spectrum above that used for the normal PON traffic. Again, the radio signals can be analog, although digitized radio signals might also be transported this way.

The analog formats are the traditional ones for RoF, but there can be problems with time-out of wireless MAC-level protocols that were not designed for long transmission delays associated with the separation of BS signal processing from the power amplifier and antenna in the field. For this reason, future designs for wireless–PON systems have the option of a radio *and* fiber (R&F) architecture [Maier], with wireless protocols restricted to the wireless domain and optical protocols to the optical domains, a solution consistent with Figure 4.18a. Figure 4.19 illustrates what an R&F system might look like. However, better performance and efficiencies could result from the development of integrated MAC-level protocols for wireless/optical networks, consistent with the parameters, delay, and others, for such integrated networks.

Table 4.1 summarizes the relative capabilities of in-PON and overlay integration. In-PON enables fast extension of the existing PON with wireless services, while overlay provides more bandwidth and supports analog radio signals. From the service provider's point of view, centralized management is desired for both modes to make the best use of optical–wireless integration. The efficiency of optical–wireless integration is enhanced by dynamic bandwidth allocation for both upstream and downstream transmissions, service mobility provisioning, network throughput enhancement, and supporting the differentiated services (DiffServ) of Internet protocol (IP) networks and the realization of a high QoS.

Mobile backhaul application of PON is motivated by the cost saving on backhaul facilities and extended coverage of mobile networks. It is efficient to share the PON infrastructure with mobile backhaul networks when BSs are located in a PON system serving area. This eliminates the requirement of

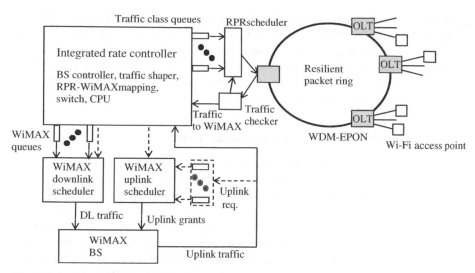

Figure 4.19 Radio and fiber, integrating an optical RPR-EPON with both WiMAX and Wi-Fi networks. Protocol processing occurs in the integrated rate controller at the WiMAX–RPR interface. RPR, resilient packet ring. Figure adapted from [Maier].

TABLE 4.1   Comparison of Optical–Wireless Integration Modes

Parameter	In-PON Integration	Overlay Integration
Wired user bandwidth	Reduced	Maintained
Wireless data unit	Encapsulated	Independent
Wireless data rate	Upper bounded by PON capacity	Independent
Extra transmission channel requirement	No	Yes

deploying a dedicated backhaul network. Wireless technologies require precise time information to support various mobility services. Ordinary PON does not provide time synchronization. Therefore, PON time synchronization is a critical challenge, and a proper solution is desired to tackle this issue. In [Luo1], the OLT plays a master role in associating the network time with the OLT local counter and distributing time to ONUs. After conducting EPON link asymmetry compensation, the OLT informs ONUs of the association between the accurate network time and ONU local counter values. An ONU is able to adjust its local time after receiving this information.

This solution was accepted by standard bodies. The IEEE 802.1AS standard [8021AS] for "Timing and Synchronization for Time-Sensitive Applications in Bridged Local Area Networks" was published in March 2011. Clause 13 addresses EPON time synchronization. The same solution is applicable to 10GE-PON [10GEPON]. The ITU-T G.984.3 Amendment 2 [9843Amd2] was published in September 2009. It specifies distributing accurate time over GPON by following the same solution. It is reused in XG-PON [XGPON] with minor extension. With these standards in place, TDMA-PONs can play their part in time distribution networks.

4.5.3 Support of Next-Generation Cellular Mobile

In order to realize very high per-user bandwidths in densely populated areas, future generations of cellular mobile communications, beginning with long term evolution-advanced (LTE-A), a fourth-generation (4G) system proposed by the 3rd Generation Partnership Project (3GPP), are expected to result in a first ITU standard in 2011. In order to realize peak data rates as high as 1 Gbps, the draft standard incorporates a technique called cooperative multipoint (CoMP), in which a distributed broadband wireless system (DBWS) of cooperating BSs rests on a high-speed optical network. PON is an excellent candidate for this supporting optical backbone.

A European project, FUTON [Monteiro], was underway at the time of writing to test the concept of optical–wireless integration to support CoMP. Figure 4.20 illustrates the FUTON high-level architecture. The DBWS in this example is joined to a central processing unit, where signal processing and coordination is done, by a CWDM-PON, in which each wavelength serves a remote access unit (RAU, a simplified BS). Each individual IF radio channel from a RAU is modulated (RoF) on a subcarrier of the wavelength carrier. In addition to supporting CoMP in 4G cellular mobile, the optical network can also support WiMAX (itself now considered a 4G technology) and legacy systems including 3G and Wi-Fi.

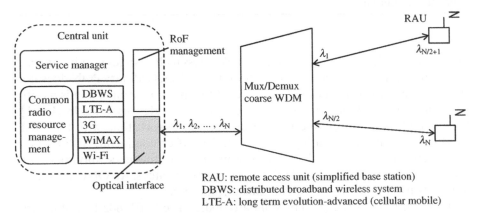

Figure 4.20 The FUTON system for support of CoMP in 4G cellular mobile communications. The figure is derived from [Monteiro].

The FUTON system, whose performance has been demonstrated by researchers in Germany [Diehm], illustrates how tightly integrated high-speed wireless communication is with high-speed optical communication. It is no longer, if it ever was, possible to regard optical and wireless as two different communication systems.

4.5.4 The Future of Optical–Wireless Integration

Going beyond traditional access systems, optical–wireless integration merges heterogeneous network technologies in a cost-effective and mutually beneficial way. We can fairly claim that the integrated architectures described above possess the following technological and operational advantages:

- higher user penetration in both urban areas desiring the flexibility of wireless access together with the capacity of optical access and rural areas where wireless may be the only practical "last mile" access technology;
- compatible with FTTx (fiber-to-the-curb, -home, etc.) deployment;
- more wired and wireless access options;
- service protection between wired and wireless segments;
- easing of the planning of wireless frequency reuse by shifting it from multiple APs to one edge node;
- colocation of radio processing and optical processing functions in edge nodes results in less expensive BSs and operating efficiencies; and
- centralized management enhances spectrum utilization, mutual protection, and operational efficiencies, and can allocate resources dynamically between the wireless and optical segments.

Further advances in optical–wireless integration will be made at different protocol layers. In the physical layer, the exploration of new transport and modulation technologies can further enhance the integrated optical–wireless network capacity. One promising approach is the combination of OFDM (downstream) and orthogonal frequency division multiple access (OFDMA) (upstream) radio signals with WDM-PON [Cvijetic]. The OFDM and OFDMA subchannels offer great service flexibility, and each subchannel can also be segmented into time slots for multiplexing in that dimension as well. Because WDM-PON provides independence for both data rate and data format, its integration with OFDM radio systems, such as WiMAX, is a natural extension to the wireless field. In the MAC layer, aggregation of control signaling has the potential to reduce transmission overhead. For example, the GPON and WiMAX standards define their own MAC layer data units to support point-to-multipoint access. They follow similar bandwidth negotiation procedures to facilitate dynamic resource allocation. Without signaling aggregation, each WiMAX connection requires a dedicated signaling process to independently negotiate its real-time transmission to the integrated edge node. Since one subscriber can initiate multiple WiMAX connections to the edge node, multiple signaling processes may exist simultaneously at one subscriber. Signaling aggregation gathers the control signals from multiple subscribers and transmits them in batches using a minimal number of GPON frames. It reduces signaling overhead by merging the overlapped processes of different signaling protocols. The challenges include how to match a WiMAX control frame to a PON control frame and how to determine the frame duration for the WiMAX data transmission.

In the management layer, it would be advantageous to realize consistent management of the integrated optical–wireless channel. For example, a uniform definition of QoS is required to consolidate the service provisioning through the optical and wireless domains. One might consider the "QoS translation" issues among the IP networking, wireless air interface, and optical transmission domains. For example, the DiffServ traffic classes must be translated into different priority treatments in the air interface, which in turn must be translated into different capacity granularities in the wavelengths, subcarrier channels, and frames of the optical segment. A unified admission control policy across optical–wireless integration and a consistent queue management scheme at the subscriber side are closely related issues.

In order to evaluate the feasibility of optical–wireless integration, more accurate cost models are needed to reveal the trade-offs among diverse integration strategies. The cost-effectiveness of optical–wireless integration depends on numerous factors, including, but not limited to, fiber trench cost, wireless tower and BS cost, installation cost, real estate cost of placing the equipment, powering cost of the active components, and spectrum license cost.

4.6 SCALING UP PON TO MUCH HIGHER TRANSMISSION RATES

A 10-Gbps PON is, at the time of writing, the high-speed state of the art for standardized, deployable PON. However, there is already an interest in the next wave of PON systems with rates beyond 10 Gbps. This is driven by the extravagantly bandwidth-hungry media applications of the near future.

For optical transmission systems, 40 and 100 Gbps are already standardized, including IEEE 802.3ba, ratified June, 2010. IEEE 802.3ba "addresses critical challenges facing technology providers today, such as the growing number of applications with demonstrated bandwidth needs far exceeding existing Ethernet capabilities, by providing a larger, more durable bandwidth pipeline. Furthermore, collaboration between the IEEE P802.3ba 40 and 100Gbps Ethernet Task Force and [ITU-T] Study Group 15 ensures these new Ethernet rates are transportable over optical transport networks" (http://standards.ieee.org/news/2010/ratification8023ba.html). IEEE 802.3ba specifies the data link layer interfaces that high-speed routers, for example, would use to pass very high-speed IP packet streams into optical Ethernet

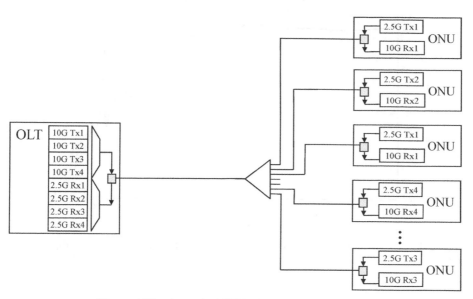

Figure 4.21   A stacked TDMA-PON with 40 Gbps.

networks, with the 100-Gbps rate focused mainly on switching in core backbone networks.

Although still in the experimental stage, 40- and 100-Gbps PONs appear feasible in several different implementations. The first approach is a backward compatible solution. It stacks a few 10-Gbps PONs (e.g., 10GE-PONs or XG-PONs) together via WDM technologies. With four pairs of wavelengths, the system is able to provide a capacity of 40 Gbps downstream and 10 Gbps upstream. As shown in Figure 4.21, this approach reuses the deployed optical distribution network (ODN). With appropriate wavelength plans, it provides coexistence with the deployed 1- and 10-Gbps PONs. Its backward compatibility with TDMA-PONs also implies cost savings to operators.

A second approach is a disruptive ODN solution. It changes the TDMA-PON ODN by adding new devices or replacing the deployed splitter. The new ODN could be an AWG, a hybrid node with a splitter and a WDM multiplexer, or an ODN with multiple stages. A recent study [Cho] suggests "to utilize the colorless light source such as the reflective semiconductor optical amplifier (RSOA)" to generate each 25-Gbps optical line signal. With CWDM, for example, four wavelengths each carrying traffic at about 25 Gbps, this system provides a rate of 100 Gbps. Even with an improved parasitic-minimizing physical structure described in this study, the nominal bandwidth of the directly modulated RSOA was only 3.2 GHz, which would normally support binary signals at only 6 Gbps. However, the roll-off of the frequency response of the improved structure is a modest 20 dB/decade, as shown in Figure 4.22, and the RSOA frequency characteristic is highly linear. This makes it possible, with aggressive electronic equalization and forward error correction (FEC)

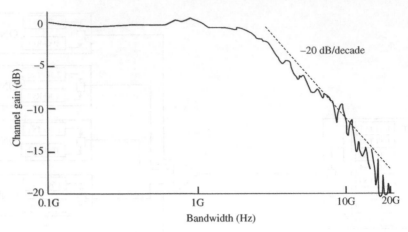

Figure 4.22   Channel gain characteristic, including RSOA, used for 25 Gbps (from [Cho]).

techniques, to realize reliable 25.78-Gbps transmission on each wavelength in the 20-nm wavelength region. Chromatic dispersion could be a problem, arising in the single-mode fiber, but can be countered with a dispersion-compensating unit in the OLT.

The third approach is an ambitious solution. It employs advanced signal processing technologies to achieve high capacity. Examples are orthogonal frequency division multiple access PON (OFDMA-PON) and coherent WDM-PON. As already noted in Section 4.3, OFDMA-PON improves multiplexing gain within a variable-width spectrum via the technologies of subcarrier communications, advanced coding, and digital signal processing (DSP). Coherent WDM-PON is based on the signal coherence between the OLT and ONUs. Local oscillators and DSP chips are used to provide dedicated wavelength channels between the OLT and each ONU. The OFDMA-PON in Figure 4.23 suppots a total capacity of 100 Gbps, at least downstream [Qian]. It avoids expensive coherent detectors by incorporating polarization multiplexing— using the same carrier frequency in orthogonally polarized signals—and direct detection.

Early implementations such as this one use electronic OFDM, which is relatively expensive for the very high rates involved, but other experiments have demonstrated direct optical generation of OFDM optical signals, which could help reduce the cost in the future.

All in all, the limited geographic extent of the typical PON appears to allow pushing transmission speeds much faster than one would normally think reasonable. It remains to be seen, and perhaps described in a future edition of this book, how soon 40- and 100-Gbps PONs will become technically and economically practical.

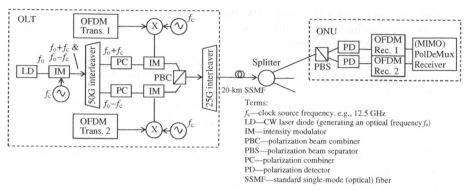

Figure 4.23 Experimental implementation of 108-Gbps OFDMA-PON using polarization multiplexing (derived from [Qian]). MIMO, multiple input multiple output; CW, coarse wavelength.

4.7 CONCLUSION

Whether you have read this through or consulted it on particular topics, we hope that you found it informative and useful. This book does not claim to be comprehensive, but we welcome comments on omissions or errors and will do our best to incorporate them in future editions. Thank you for joining us in this fascinating area of communications technology and services.

The Authors.

REFERENCES

[10GEPON] IEEE Std 802.3av-2009, "Physical layer specifications and management parameters for 10 Gb/s passive optical networks," IEEE, 2009.

[10GEPONINTEROP] "ZTE hosts IEEE 10G-interoperability showcase," available at http://wwwen.zte.com.cn/en/press_center/news/201004/t20100422_183446.html

[8021AS] IEEE Std 802.1AS-2011, "Timing and synchronization for time-sensitive applications in bridged local area networks," IEEE, 2011.

[9843Amd2] ITU-T G.984.3 Recommendation Amendment 2, "Gigabit-capable passive optical networks (G-PON): Transmission convergence layer specifications," ITU-T, 2009.

[Banerjee] A. Banerjee, Y. Park, F. Clarke, H. Song, & S. Yang, "Wavelength-division-multiplexed passive optical network (WDM-PON) technologies for broadband access: A review," J. Optical Netw., 4(11), November, 2005.

[Beyranvan] H. Beyranvan & J. Salehi, "Multirate and multi-quality-of-service passive optical network based on hybrid WDM/OCM system," IEEE Communications Magazine, February, 2011.

[Cheng] N. Cheng and F. Effenberger, "WDM PON: Systems and technologies," ECOC Workshop, Torino, Italy, 2010, available at http://www.ecoc2010.org/contents/attached/c20/WS_5_Cheng.pdf

[Cho] K. Cho, B. Choi, Y. Takushima and Y. Chung, "25.78-Gb/s operation of RSOA for next-generation optical access networks," IEEE Photonics Technol. Lett., 23(8), April 15, 2011.

[Cioffi et al.] J.M. Cioffi, S. Jagannathan, M. Mohseni, & G. Ginis, "CuPON: The copper alternative to PON 100 GB/s DSL networks," IEEE Communications Magazine, June, 2007.

[Cvijetic] N. Cvijetic, D. Qian, T. Wang, & S. Weinstein, "OFDM for next-generation optical access networks," Proc. IEEE Wireless-Optical Workshop (held in conjunction with IEEE WCNC), Sydney, Australia, April, 2010.

[Darcie] T.E. Darcie & G.E. Bodeep, "Lightwave subcarrier CATV transmission systems," IEEE Trans. Microwave Theory Tech., 38(5), pp. 524–533, May 1990.

[Davey] R. Davey, P. Healey, I. Hope, P. Watkinson, D. Payne, O. Marmur, J. Ruhmann, & Y. Zuiderveld, "DWDM reach extension of a GPON to 135 km," J. Lightwave Technol., 24(1), pp. 29–31, January 2006.

[Diehm] F. Diehm, P. Marsch, & G. Fettweis, "The FUTON prototype: Proof of concept for coordinated multi-point in conjunction with a novel integrated wireless/optical architecture," Proc. IEEE Wireless-Optical Workshop (held in conjunction with IEEE WCNC), Sydney, Australia, April, 2010.

[DXW] I. Djordjevic, L. Xu, & T. Wang, "Simultaneous chromatic dispersion and PMD compensation by using coded-OFDM and girth-10 LDPC codes," OSA Optics Expr., 16(14), pp. 10269–10278, July, 2008.

[Fabry–Perot] Available at http://www.fiber-optics.info/articles/laser-diodes.

[Green] (already cited in an earlier chapter).

[Grobe] K. Grobe, M. Roppelt, A. Autenrieth, J.-P. Elbers, & M. Eiselt, "Cost and energy consumption analysis of advanced WDM-PONs," IEEE Communications Magazine, February, 2011.

[Hanzo et al.] L. Hanzo, M. Munster, B. Choi, & T. Keller, OFDM and MC-CDMA for Broadband Multi-user Communications, WLANs and Broadcasting, Wiley, 2003.

[ITU Plugtest 2010] "ITU-T Plugtest for fault/error recovery behavior and QoS," available at http://www.etsi.org/plugtests/GPON12/GPON.htm.

[Kamei] S. Kamei, Y. Inoue, T. Shibata, & A. Kaneko, "Low-loss and compact silica-based athermal arrayed waveguide grating using resin-filled groove," IEEE/OSA J. Lightwave Technol., 27(17), September 2009.

[Kuroda] T. Kuroda, "Current status on super HDTV development in Japan," Proc. 2010 IEEE International Symposium on Broadband Multimedia Systems and Broadcasting, Shanghai, China, March 24–26, 2010.

[Lam] C. Lam, Passive Optical Networks: Principles and Practice, Academic Press, 2007.

[LeeMun] C.H. Lee & S.G. Mun, "WDM-PON based on wavelength locked Fabry-Perot LDs," J. Optical Society of Korea, 12(4), pp. 326–336, December, 2008.

[LeeMunLee] S.-M. Lee, S.-G. Mun, & C.-H. Lee, "Consolidation of a metro network into an access network based on long-reach DWDM-PON," Proc. Optical Fiber Communication Conf., March, 2006.

[Lightwave2004] "FSAN achieves BPON voice interoperability," available at http://www.lightwaveonline.com/business/news/53909677.html.

[Luo1] Y. Luo, F. Effenberger, and N. Ansari, "Time synchronization over Ethernet passive optical networks," to appear in IEEE Communications Magazine, 2012.

[Maier] M. Maier & N. Ghazisaidi, "QoS-aware radio-and-fiber (R&F) access-metro networks," Proc. IEEE Wireless-Optical Workshop (held in conjunction with IEEE WCNC), Sydney, Australia, April, 2010.

[Monteiro] P. Monteiro, E. Lopez, D. Wake, N. Gomes, & A. Gameiro, "Fiber optic networks for distributed radio architectures: FUTON concept and operation," Proc. IEEE Wireless-Optical Workshop (held in conjunction with IEEE WCNC), Sydney, Australia, April, 2010.

[Pinter1] S. Pinter & X. Fernando, "Fiber-wireless solution for broadband multimedia access," IEEE Canadian Review, First Quarter 2005, available at http://www.ee.ryerson.ca/wincore/ADROIT/IEEE_CR_final.pdf

[Pinter2] S. Pinter & X. Fernando, "Estimation and equalization of fiber-wireless uplink for multiuser CDMA 4G networks," IEEE Trans. on Communications, 58(6), pp. 1803–1813, June 2010.

[Qian] D. Qian, N. Cvijetic, J. Hu, & T. Wang, "108 Gb/s OFDMA-PON with polarization multiplexing and direct detection," IEEE J. Lightwave Technol., 28(4), pp. 484–493, February, 2010.

[Salim] J. Salim, U.S. Patent 6,229,632, "Broadband optical transmissions system utilizing differential wavelength modulation," 2001.

[Shibutani_et al.] M. Shibutani, W. Domon, & K. Emura, "Optical multiple-access network with subcarrier multiplexing," Proc. IEEE Optical/Hybrid Access Networks 1993.

[Shin] D. Shin, D. Jung, J.K. Lee, J.H. Le, Y. Choi, Y. Bang, H. Shing, J. Lee, S. Hwang, & Y. Oh, "155 Mbit/s transmission using ASE-injected Fabry–Perot laser diode in WDM-PON over 70/spl deg/C temperature range," IEEE Electron. Lett., 39(18), pp. 1331–1332, September 4, 2003.

[SARDANA1] J. Prat, "SARDANA project report to future networks concertation meeting," Brussels, 11–12 March, 2008, available at http://ftp.cordis.europa.eu/pub/fp7/ict/docs/future-networks/projects-sardana-concertation200803g_en.pdf

[SARDANA2] J. Segarra, J. Part, P. Chanclou, R. Soila, S. Spirou, A. Teixeira, G. Tosi-beleffi, and I. Tomkos, "SARDANA: An all-optical access-metro WDM/TDM-PON," slides shown at 49th FITCE (Federation of Telecommunications Engineers of the European Union) Congress, Santiago de Compostela, Spain, September, 2010, available at http://www.fitce2010.org/ponencias/1_VIERNES_SESION6_Josep_Segarra.pdf

[Teixeira] A. Teixeira, A. Batista, N. Pavlovic, P. Andre, D. Forin, & G. Beleffi, "Analysis of remote nodes parameters in passive optical networks," 5th Luzo-Mozambican Engineering Conf., Maputo, September 2–4, 2008.

[Telcordia GPON 2006] "ITU G-PON Pavilion @ Telecom world 2006," Telcordia report prepared for Q2/SG15 Interim Meeting, February 9, 2007, available at http://www.telcordia.com/services/testing/integrated-access/2006g-pon.pdf.

[Telcordia GPON 2008] "ITU G-PON Pavilion @ NXTcomm 2008," available at http://www.telcordia.com/services/testing/integrated-access/2008g-pon_overview.pdf

[TMCnet EFM 2004] "Ethernet in the first mile alliance announces official ratification of the IEEE 801.3ah EFM standard," TMCnet, June 24, 2004, available at http://www.tmcnet.com/usubmit/2004/jun/1052016.htm

[XGPON] ITU-T G.987.3 Recommendation, 10-Gigabit-capable passive optical networks (XG-PON): Transmission convergence (TC) layer specifications, ITU-T, 2010.

[Yu1] J. Yu, O. Akanbi, Y. Luo, L. Zong, Z. Jia, T. Wang, & G.-K. Chang, "A novel WDM-PON architecture with centralized lightwaves in the OLT for providing triple play services," Proc. IEEE/OSA Optical Fiber Conference 2007.

[Yu2] J. Yu, G.-K. Chang, T. Koonen, and G. Ellinas, "Radio-over-optical-fiber networks," J. Opt. Networks, 7, numbers iii–iv, 2008.

APPENDIX: EXCERPTS FROM THE IEEE 10 Gbps EPON STANDARD 802.3av-2009

Among the major passive optical network (PON) standards, Institute of Electrical and Electronics Engineers (IEEE) 802.3av (10-Gbps Ethernet) holds particular promise for economical broadband access that is convenient for high-speed data communications, high-definition video programming, real-time communications, and many other business and residential applications. The following excerpts from the 236-page document demonstrate the style and structure of an important standard and some parts of the critical architectural description, focusing on the new physical (PHY) and medium access control (MAC) elements. The standard includes considerable detail on signaling interfaces, control, signal design constraints, and testing, all important to interoperability, while leaving a great deal of functionality, such as dynamic bandwidth assignment, to proprietary implementations by the operators deploying PON systems. The standard includes a series of compliance questionnaire forms from which we draw a couple of samples.

We include the full table of contents but omit other introductory sections including the list of participants in the working group that developed the standard and description of the IEEE's rights and intentions regarding the use of the standard.

This standard is not a complete specification on its own. Its components are largely changes to sections of IEEE Std 802.3-2008, the primary Ethernet standard. References made to particular sections or tables are directed to the 802.3 standard.

This standard is copyright 2009 by the IEEE, with all rights reserved.

The ComSoc Guide to Passive Optical Networks: Enhancing the Last Mile Access,
First Edition. Stephen Weinstein, Yuanqiu Luo, Ting Wang.
© 2012 Institute of Electrical and Electronics Engineers. Published 2012 by John Wiley & Sons, Inc.

Excerpts from the standard begin below. Omitted portions of the standard are indicated by a dashed line.

IEEE STANDARD FOR INFORMATION TECHNOLOGY— TELECOMMUNICATIONS AND INFORMATION EXCHANGE BETWEEN SYSTEMS—LOCAL AND METROPOLITAN AREA NETWORKS—SPECIFIC REQUIREMENTS

Part 3: Carrier Sense Multiple Access with Collision Detection (CSMA/CD) Access Method and Physical Layer Specifications

Amendment 1: Physical Layer Specifications and Management Parameters for 10 Gb/s Passive Optical Networks

IEEE Computer Society
Sponsored by the LAN/MAN Standards Committee
IEEE Std 802.3av™ - 2009 (amendment of IEEE Std 802.3™ - 2008), 30 October 2009

- - - - - - - -

Abstract: This amendment to IEEE Std 802.3-2008 extends Ethernet Passive Optical Networks (EPONs) operation to 10 Gb/s providing both symmetric, 10 Gb/s downstream and upstream, and asymmetric, 10 Gb/s downstream and 1 Gb/s upstream, data rates. It specifies the 10 Gb/s EPON Reconciliation Sublayer, 10GBASE-PR symmetric and 10/1GBASE-PRX Physical Coding Sublayers (PCSs) and Physical Media Attachments (PMAs), and Physical Medium Dependent sublayers (PMDs) that support both symmetric and asymmetric data rates while maintaining complete backward compatibility with already deployed 1 Gb/s EPON equipment. The EPON operation is defined for distances of at least 10 km and at least 20 km, and for split ratios of 1:16 and 1:32.

An additional MAC Control opcode is also defined to provide organization specific extension operation.

Keywords: 10 Gb/s Ethernet Passive Optical Networks (10G-EPON), forward error correction (FEC), Multi-Point MAC Control (MPMC), Physical Coding Sublayer (PCS), Physical Media Attachment (PMA), Physical Medium Dependent (PMD), PON, Point to Multipoint (P2MP), Reconciliation Sublayer (RS)

- - - - - - - -

INTRODUCTION (Not officially part of the 802.3av standard)

IEEE Std 802.3™ was first published in 1985. Since the initial publication, many projects have added functionality or provided maintenance updates to the specifications and text included in the standard. Each IEEE 802.3 project/amendment is identified with a suffix (e.g., IEEE Std 802.3av-2009).

The Media Access Control (MAC) protocol specified in IEEE Std 802.3 is Carrier Sense Multiple Access with Collision Detection (CSMA/CD). This MAC protocol was included in the experimental Ethernet developed at Xerox Palo Alto Research Center. While the experimental Ethernet had a 2.94 Mb/s data rate, IEEE Std 802.3-1985 specified operation at 10 Mb/s. Since 1985 new media options, new speeds of operation, and new capabilities have been added to IEEE Std 802.3.

Some of the major additions to IEEE Std 802.3 are identified in the marketplace with their project number. This is most common for projects adding higher speeds of operation or new protocols. For example, IEEE Std 802.3u™ added 100 Mb/s operation (also called Fast Ethernet), IEEE Std 802.3x™ specified full duplex operation and a flow control protocol, IEEE Std 802.3z™ added 1000 Mb/s operation (also called Gigabit Ethernet), IEEE Std 802.3ae™ added 10 Gb/s operation (also called 10 Gigabit Ethernet) and IEEE Std 802.3ah™ specified access network Ethernet (also called Ethernet in the First Mile). These major additions are all now included in, and are superseded by, IEEE Std 802.3-2008 and are not maintained as separate documents.

At the date of IEEE Std 802.3av-2009 publication, IEEE Std 802.3 is comprised of the following documents:

IEEE Std 802.3-2008

Section One—Includes Clause 1 through Clause 20 and Annex A through Annex H and Annex 4A. Section One includes the specifications for 10 Mb/s operation and the MAC, frame formats and service interfaces used for all speeds of operation.

Section Two—Includes Clause 21 through Clause 33 and Annex 22A through Annex 33E. Section Two includes management attributes for multiple protocols and speed of operation as well as specifications for providing power over twisted pair cabling for multiple operational speeds. It also includes general information on 100 Mb/s operation as well as most of the 100 Mb/s Physical Layer specifications.

Section Three—Includes Clause 34 through Clause 43 and Annex 36A through Annex 43C. Section Three includes general information on 1000 Mb/s operation as well as most of the 1000 Mb/s Physical Layer specifications.

Section Four—Includes Clause 44 through Clause 55 and Annex 44A through Annex 55B. Section Four includes general information on 10 Gb/s operation as well as most of the 10 Gb/s Physical Layer specifications.

Section Five—Includes Clause 56 through Clause 74 and Annex 57A through Annex 74A. Clause 56 through Clause 67 and associated annexes specify subscriber access and other Physical Layers and sublayers for operation from 512 kb/s to 1000 Mb/s, and defines services and protocol elements that enable the exchange of IEEE Std 802.3 format frames between stations in a subscriber access network. Clause 68 specifies a 10 Gb/s Physical Layer specification.

Clause 69 through Clause 74 and associated annexes specify Ethernet operation over electrical backplanes at speeds of 1000 Mb/s and 10 Gb/s.

IEEE Std 802.3av-2009

This amendment includes changes to IEEE Std 802.3-2008 and adds Clause 75 through Clause 77 and Annex 75A through Annex 76A. This amendment adds new Physical Layers for 10 Gb/s operation on point-to-multipoint passive optical networks.

IEEE Std 802.3bc™-2009

This amendment includes changes to IEEE Std 802.3-2008 and adds Clause 79. This amendment moves the Ethernet Organizationally Specific Type, Length, Value (TLV) information elements that were specified in IEEE Std 802.1AB to IEEE Std 802.3.

IEEE Std 802.3at™-2009

This amendment includes changes to IEEE Std 802.3-2008. This amendment augments the capabilities of IEEE Std 802.3-2008 with higher power levels and improved power management information.

IEEE Std 802.3 will continue to evolve. New Ethernet capabilities are anticipated to be added within the next few years as amendments to this standard.

- - - - - - - -

Contents [page numbers at right are those in the IEEE 802.3av standard]

- - - - - - - -
Annex 31A
(normative)

MAC Control opcode assignments

Change Table 31A.1 as follows:

TABLE 31A.1   MAC Control Opcodes

Opcode (hexadecimal)	MAC Control function	Specified in	Value/Comment	Time stamp[a]
00-00	Reserved			
00-01	PAUSE	Annex 31B	Requests that the recipient stop transmitting non-control frames for a period of time indicated by the parameters of this function.	No
00-02	GATE	Clause 64 Clause 77	Request that the recipient allow transmission of frames at a time, and for a period of time indicated by the parameters of this function.	Yes
00-03	REPORT	Clause 64 Clause 77	Notify the recipient of pending transmission requests as indicated by the parameters of this function.	Yes
00-04	REGISTER_REQ	Clause 64 Clause 77	Request that the station be recognized by the protocol as participating in a gated transmission procedure as indicated by the parameters of this function.	Yes
00-05	REGISTER	Clause 64 Clause 77	Notify the recipient that the station is recognized by the protocol as participating in a gated transmission procedure as indicated by the parameters of this function.	Yes
00-06	REFISTER_ACK	Clause 64 Clause 77	Notify the recipient that the station acknowledges participation in a gated transmission procedure.	Yes
00-07 through FF-FFD	Reserved			

Opcode (hexadecimal)	MAC Control function	Specified in	Value/Comment	Time stamp[a]
FF-FE	EXTENSION	Annex 31C	This frame is used for Organization–Specific Extension. Upon reception of this message the MAC Control generates MA CONTROL Indication informing the MAC Control Client to perform the relevant action.	No
FF-FF	Reserved			

[a] The timestamp field is generated by MAC Control and is not exposed through the client interface.

Insert a new row to Table 31A.3 at the end of the table, as presented below:

TABLE 31A.3   GATE MAC Control Indications

Discovery information[a]	16 bits	See Table 77–3 for the internal structure of the discoveryInformation field.

[a] Only present in 10G–EPON GATE MAC Control indication (Clause 77).

Insert three new rows to Table 31A.5 at the end of the table, as presented below:

TABLE 31A.5   REGISTER_REQ MAC Control Indications

Laser on time[a]	8 bits	Indicates the Laser On Time characteristic for the given ONU transmitter, expressed in the units of time_quanta.
Laser off time[a]	8 bits	Indicates the Laser Off Time characteristic for the given ONU transmitter, expressed in the units of time_quanta.
Discovery information[a]	16 bits	See Table 77–7 for the internal structure of the discoveryInformation field.

[a] Only present in 10G–EPON REGISTER_REQ MAC Control indication (Clause 77).

Insert two new rows to Table 31A.6 at the end of the table, as presented below:

TABLE 31A.6   REGISTER MAC Control Indications

Laser on time[a]	8 bits	Indicates the Laser On Time characteristic for the given ONU transmitter, expressed in the units of time_quanta.
Laser off time[a]	8 bits	Indicates the Laser Off Time characteristic for the given ONU transmitter, expressed in the units of time_quanta.

[a] Only present in 10G–EPON REGISTER MAC Control indication (Clause 77).

Add a new Table 31A.8 after Table 31A.7 with the description of EXTEN-
SION frame with the following contents:

TABLE 31A.8   EXTENSION MAC Control Indications

EXTENSION (opcode oxFFFE)		
Indication_operand_ list element	Value	Interpretation
OUI	24 bits	Organizationally-Unique Identifier that determines the format and semantics of the Value field and its subfields, if any are defined.
Value	Variable	

- - - - - - - -

Insert the following new clauses and corresponding annexes after Clause 74:

75. Physical Medium Dependent (PMD) sublayer and medium for passive optical networks, type 10GBASE–PR and 10/1GBASE–PRX

75.1 OVERVIEW

Clause 75 describes Physical Medium dependent (PMD) sublayer for Ethernet
Passive Optical Networks operating at the line rate of 10.3125 GBd in either
downstream or in both downstream and upstream directions.

75.1.1 Terminology and Conventions

The following list contains references to terminology and conventions used in
Clause 75:

Basic terminology and conventions, see 1.1 and 1.2.

Normative references, see 1.3.

Definitions, see 1.4.

Abbreviations, see 1.5.

Informative references, see Annex A.

Introduction to 1000 Mb/s baseband networks, see Clause 34.

Introduction to 10 Gb/s baseband network, see Clause 44.

Introduction to Ethernet for subscriber access networks, see Clause 56.

EPONs operate over a point-to-multipoint (P2MP) topology, also called a tree
or trunk-and-branch topology. The device connected at the root of the tree is
called an Optical Line Terminal (OLT) and the devices connected as the leaves
are referred to as Optical network Units (ONUs). The direction of transmission from the OLT to the ONUs is referred to as the downstream direction,
while the direction of transmission from the ONUs to the OLT is referred to
as the upstream direction.

75.1.2 Goals and Objectives

The following are the PMD objectives fulfilled by Clause 75:

a. Support subscriber access networks using point-to-multipoint topologies on optical fiber.

b. Provide physical layer specifications:

1. PHY for PON, 10 Gb/s downstream/1 Gb/s upstream, on a single SMF

2. PHY for PON, 10 Gb/s downstream/10 Gb/s upstream, on a single SMF

c. PHY(s) to have a BER better than or equal to 10–12 at the PHY service interface.

d. Define up to three optical power budgets that support split ratios of at least 1:16 and at least 1:32, and distances of at least 10 km and at least 20 km.

75.1.3 Power Budget Classes

To support the above-stated objectives, Clause 75 defines the following three power budget classes:

- *Low power budget class* supports P2MP media channel insertion loss of ≤20 dB e.g., a PON with the split ratio of at least 1:16 and the distance of at least 10 km.

- Medium power budget class supports P2MP media channel insertion loss of ≤24 dB e.g., a PON with the split ratio of at least 1:16 and the distance of at least 20 km or a PON with the split ratio of at least 1:32 and the distance of at least 10 km.

- High power budget class supports P2MP media channel insertion loss of ≤29 dB e.g., a PON with the split ratio of at least 1:32 and the distance of at least 20 km.

75.1.4 Power Budgets

Each power budget class is represented by PRX-type power budget and PR-type power budget as follows:

- PRX–type power budget describes asymmetric-rate PHY for PON operating at 10 Gb/s downstream and 1 Gb/s upstream over a single SMF [see objective b 1) in 75.1.2].

- PR–type power budget describes symmetric-rate PHY for PON operating at 10 Gb/s downstream and 10 Gb/s upstream over a single SMF [see objective b 2) in 75.1.2].

Each power budget is further identified with a numeric representation of its class, where a value of 10 represents low power budget, a value of 20 represents medium power budget, and a value of 30 represents high power budget. Thus, the following power budgets are defined in Clause 75:

- *PRX10*: asymmetric-rate, low power budget, compatible with PX10 power budget defined in Clause 60.
- *PRX20*: asymmetric-rate, medium power budget, compatible with PX20 power budget defined in Clause 60.
- *PRX30*: asymmetric-rate, high power budget.
- *PR10*: symmetric-rate, low power budget, compatible with PX10 power budget defined in Clause 60.
- *PR20*: symmetric-rate, medium power budget, compatible with PX20 power budget defined in Clause 60.
- *PR30*: symmetric-rate, high power budget.

Table 75.1 shows the primary attributes of all power budget types defined in Clause 75.

TABLE 75.1   Power Budgets

Description	Low Power Budget		Medium Power Budget		High Power Budget		Units
	PRX10	PR10	PRX20	PR20	PRX30	PR30	
Number of fibers	1						–
Nominal downstream line rate	10.3125						GBd
Nominal upstream line rate	1.25	10.3125	1.25	10.3125	1.25	10.3125	GBd
Nominal downstream wavelength	1577						nm
Downstream wavelength tolerance	−2, +3						nm
Nominal upstream wavelength	1310	1270	1310	1270	1310	1270	nm
Upstream wavelength tolerance	±50	±10	±50	±10	±50	±10	nm
Maximum reach[a]	≥10		≥20		≥20		km
Maximum channel insertion loss	20		24		29		dB
Minimum channel insertion loss	5		10		15		dB

[a] A compliant system may exceed the maximum reach designed for given power budget as long as optical power budget and other mandatory optical layer specifications are met.

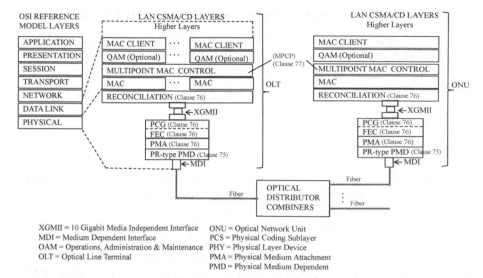

Figure 75.1 Relationship of 10/10G–EPON P2MP PMD to the ISO/IEC OSI reference model and the IEEE 802.3 CSMA/CD LAN model.

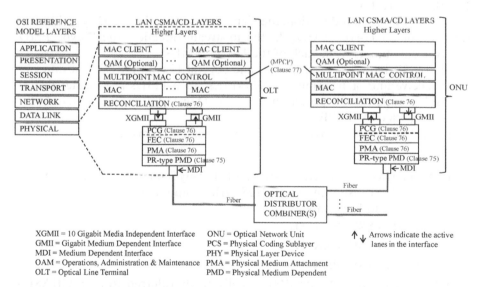

Figure 75.2 Relationship of 10/1G–EPON P2MP PMD to the ISO/IEC OSI reference model and the IEEE 802.3 CSMA/CD LAN model.

75.1.5   Positioning of PMD Sublayer within the IEEE 802.3 Architecture

Figures 75.1 and 75.2 depict the relationships of the symmetric-rate (10/10G–EPON) and asymmetric-rate (10/1G–EPON) PMD sublayer (shown hatched) with other sublayers and the ISO/IEC Open System Interconnection (OSI) reference model.

75.2   PMD TYPES

Similarly to power budget classes, asymmetric-rate and symmetric-rate PMDs are identified by PRX and PR designations, respectively.

The characteristics of the P2MP topology result in significantly different ONU and OLT PMDs. For example, the OLT PMD operates in a continuous mode in the transmit direction (downstream), but uses a burst mode in the receive direction (upstream). On the other hand, the ONU PMD receives data in a continuous mode, but transmits in a burst mode. To differentiate OLT PMDs from ONU PMDs, the OLT PMD name has a suffix "D" appended to it, where D stands for downstream–facing PMD, for example, 10GBASE–PR–D1. ONU PMDs have suffix "U" for upstream–facing PMD, for example, 10GBASE–PR–U1.

In the downstream direction, the signal transmitted by the D-type PMD is received by all U-type PMDs. In the upstream direction, the D-type PMD receives data bursts from each of the U-type PMDs.

Clause 75 defines several D-type and several U-type PMDs, that differ in their receive and/or transmit characteristics. Such PMDs are further distinguished by appending a digit after the suffix D or U, for example, 10GBASE–PR–D1 or 10GBASE–PR–D2.

The following OLT PMDs (D-type) are defined in this subclause:

a. Asymmetric-rate D-type PMDs (collectively referred to as 10/1GBASE–PRX–D), transmitting at 10.3125 GBd continuous mode and receiving at 1.25 GBd burst mode:
   1. 10/1GBASE–PRX–D1
   2. 10/1GBASE–PRX–D2
   3. 10/1GBASE–PRX–D3
b. Symmetric-rate D-type PMDs (collectively referred to as 10GBASE–PR–D), transmitting at 10.3125 GBd continuous mode and receiving at 10.3125 GBd burst mode:
   1. 10GBASE–PR–D1
   2. 10GBASE–PR–D2
   3. 10GBASE–PR–D3

The following ONU PMDs (U-type) are defined in this subclause:

c. Asymmetric-rate U-type PMDs (collectively referred to as 10/1GBASE–PRX–U), transmitting at 1.25 GBd burst mode and receiving at 10.3125 GBd continuous mode:
   1. 10/1GBASE–PRX–U1
   2. 10/1GBASE–PRX–U2
   3. 10/1GBASE–PRX–U3
d. Symmetric-rate U-type PMDs (collectively referred to as 10GBASE–PR–U), transmitting at 10.3125 GBd burst mode and receiving at 10.3125 GBd continuous mode:
   1. 10GBASE–PR–U1
   2. 10GBASE–PR–U3

A specific power budget is achieved by combining an OLT PMD (D-type) with an ONU PMD (U-type) as shown in 75.2.1. Detailed PMD receive and transmit characteristics for D-type PMDs are given in 75.4 and characteristics for U-type PMDs are presented in 75.5. Every PMD has non-overlapping transmit and receive wavelength bands and operates over a single SMF (see 75B.2).

75.2.1 Mapping of PMDs to Power Budgets

The power budget is determined by the PMDs located at the ends of the physical media. This subclause describes how PMDs may be combined to achieve the power budgets listed in Table 75.1.

75.2.1.1 Asymmetric-Rate, 10 Gb/s Downstream and 1 Gb/s Upstream power Budgets (PRX Type)   Table 75.2 illustrates recommended pairings of asymmetric-rate ONU PMDs with asymmetric-rate OLT PMDs to achieve the power budgets shown in Table 75.1.

TABLE 75.2   PMD—Power Budget Mapping for Asymmetric-rate PRX–type Power Budgets

		OLT PMDs		
		10/1GBASE–PRX–D1	10/1GBASE–PRX–D2	10/1GBASE–PRX–D3
ONU PMDs	10/1GBASE–PRX–U1	PRX10	N/A	N/A
	10/1GBASE–PRX–U2	N/A	PRX20	N/A
	10/1GBASE–PRX–U3	N/A	N/A	PRX30

**TABLE 75.3   PMD—Power Budget Mapping for Symmetric-rate
PR–type Power Budgets**

		OLT PMDs		
		10GBASE–PR–D1	10GBASE–PR–D2	10GBASE–PR–D3
ONU	10GBASE–PR–U1	PR10	PR20	N/A
PMDs	10GBASE–PR–U3	N/A	N/A	PR30

75.2.1.2   Symmetric-rate, 10 Gb/s Power Budgets (PR Type)   Table 75.3 illustrates recommended pairings of symmetric-rate ONU PMDs with symmetric-rate OLT PMDs to achieve the power budgets as shown in Table 75.1.

75.3   PMD FUNCTIONAL SPECIFICATIONS

The 10GBASE–PR and 10/1GBASE–PRX type PMDs perform the transmit and receive functions that convey data between the PMD service interface and the MDI.

75.3.1   PMD Service Interface

The following specifies the services provided by Clause 75 PMDs. These PMD sublayer service interfaces are described in an abstract manner and do not imply any particular implementation.

The PMD Service Interface supports the exchange of a continuous stream of bits, representing either 64B/66B blocks (the transmit and receive paths in 10GBASE–PR PMDs, transmit path in 10/1GBASE–PRX–D PMDs) or 8B/10B blocks (transmit path in 10/1GBASE–PRX–U PMDs, receive path in 10/1GBASE–PRX–D PMDs), between the PMA and PMD entities. The PMD translates the serialized data received from the compatible PMA to and from signals suitable for the specified medium. The following primitives are defined:

- PMD_UNITDATA.request
- PMD_UNITDATA.indication
- PMD_SIGNAL.request
- PMD_SIGNAL.indication

75.3.1.1   Delay Constraints   The PMD shall introduce a transmit delay of not more than 4 time_quanta with the variability of no more than 0.5 time_quanta, and a receive delay of not more than 4 time_quanta with the variability of no more than 0.5 time_quanta. A description of the overall system delay constraints can be found in 76.1.2 and the definition for the time_quantum can be found in 77.2.2.1.

75.3.1.2 PMD_UNITDATA.Request This primitive defines the transfer of a serial data stream from the Clause 65 or Clause 76 PMA to the PMD.

The semantics of the service primitive are PMD_UNITDATA.request(tx_bit). The data conveyed by PMD_UNITDATA.request is a continuous stream of bits. The tx_bit parameter can take one of two values: ONE or ZERO. The Clause 76 PMA continuously sends the appropriate stream of bits to the PMD for transmission on the medium, at a nominal signaling speed of 10.3125 GBd in the case of 10/10G–EPON OLT, 10/10G–EPON ONU, and 10/1G–EPON OLT PMDs. The Clause 65 PMA continuously sends the appropriate stream of bits to the PMD for transmission on the medium, at a nominal signaling speed of 1.25 GBd in the case of 10/1G–EPON ONU PMDs. Upon the receipt of this primitive, the PMD converts the specified stream of bits into the appropriate signals at the MDI.

75.3.1.3 PMD_UNITDATA.Indication This primitive defines the transfer of data from the PMD to the Clause 65 or Clause 76 PMA.

The semantics of the service primitive are PMD_UNITDATA.indication(rx_bit). The data conveyed by PMD_UNITDATA.indication is a continuous stream of bits. The rx_bit parameter can take one of two values: ONE or ZERO. The PMD continuously sends a stream of bits to the Clause 76 PMA corresponding to the signals received from the MDI, at the nominal signaling speed of 10.3125 GBd in the case of 10/10G–EPON OLT, 10/10G–EPON ONU, and 10/1G–EPON ONU PMDs or to the Clause 65 PMA at the nominal signaling speed of 1.25 GBd in the case of 10/1G–EPON OLT PMDs.

75.3.1.4 PMD_SIGNAL. Request In the upstream direction, this primitive is generated by the Clause 76 PMA to turn on and off the transmitter according to the granted time. A signal for laser control is generated as described in 76.4.1.1 for the Clause 76 PCS.

The semantics of the service primitive are PMD_SIGNAL. request(tx_enable). The tx_enable parameter can take on one of two values: ENABLE or DISABLE, determining whether the PMD transmitter is on (enabled) or off (disabled). The Clause 76 PMA generates this primitive to indicate a change in the value of tx_enable. Upon the receipt of this primitive, the PMD turns the transmitter on or off as appropriate.

75.3.1.5 PMD_SIGNAL. Indication This primitive is generated by the PMD to indicate the status of the signal being received from the MDI.

The semantics of the service primitive are PMD_SIGNAL. indication (SIGNAL_DETECT). The SIGNAL_DETECT parameter can take on one of two values: OK or FAIL, indicating whether the PMD is detecting light at the receiver (OK) or not (FAIL). When SIGNAL_DETECT = FAIL, PMD_UNITDATA.indication(rx_bit) is undefined. The PMD generates this primitive to indicate a change in the value of SIGNAL_DETECT. If the MDIO interface is implemented, then PMD_global_signal_detect shall be continuously set to the value of SIGNAL_DETECT.

NOTE—SIGNAL_DETECT = OK does not guarantee that PMD_UNIT-DATA.indication(rx_bit) is known good. It is possible for a poor quality link to provide sufficient light for a SIGNAL_DETECT = OK indication and still not meet the specified bit error ratio. PMD_SIGNAL.indication (SIGNAL_DETECT) has different characteristics for upstream and downstream links, see 75.3.5.

75.3.2 PMD Block Diagram

The PMD sublayer is defined at the eight reference points shown in Figure 75.3 for 10GBASE–PR and 10/1GBASE–PRX PMDs.

For 10GBASE–PR and 10/1GBASE–PRX PMDs, test points TP1 through TP4 refer to the downstream channel, while test points TP5 through TP8 refer to the upstream channel. In the downstream channel, TP2 and TP3 are compliance points, while in the upstream channel TP6 and TP7 are compliance points. TP1, TP4, TP5, and TP8 are reference points for use by implementers. The optical transmit signal is defined at the output end of a patch cord (TP2 for the downstream channel and TP6 for the upstream channel), between 2 m and 5 m in length, of a fiber type consistent with the link type connected to the transmitter. Unless specified otherwise, all transmitter measurements and tests defined in 75.7 are made at TP2 or TP6, while tests defined in 60.7 are made at TP6. The optical receive signal is defined at the output of the fiber optic cabling (TP3 for the downstream channel and TP7 for the upstream channel) connected to the receiver. Unless specified otherwise, all receiver measurements and tests defined in 75.7 are made at TP3 or TP7.

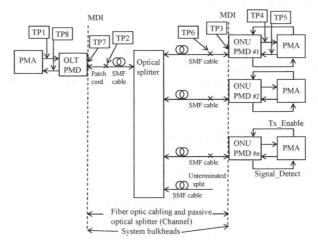

Figure 75.3   10GBASE–PR and 10/1GBASE–PRX block diagram.

The electrical specifications of the PMD service interface (TP1 and TP4 for the downstream channel and TP5 and TP8 for the upstream channel) are not system compliance points (these are not readily testable in a system implementation).

75.3.3 PMD Transmit Function

The PMD Transmit function shall convey the bits requested by the PMD service interface message PMD_UNITDATA.request (tx_bit) to the MDI according to the optical specifications in Clause 75.

In the upstream direction, the flow of bits is interrupted according to PMD_ SIGNAL. request (tx_enable). This implies three optical levels, 1, 0, and dark, the latter corresponding to the transmitter being in the OFF state. The higher optical power level shall correspond to tx_bit = ONE.

75.3.4 PMD Receive Function

The PMD Receive function shall convey the bits received from the MDI according to the optical specifications in Clause 75 to the PMD service interface using the message PMD_UNITDATA.indication (rx_bit). The higher optical power level shall correspond to rx_bit = ONE.

75.3.5 PMD Signal Detect Function

75.3.5.1 ONU PMD Signal Detect The PMD Signal Detect function for the continuous 'mode downstream signal shall report to the PMD service interface, using the message PMD_SIGNAL.indication (SIGNAL_DETECT), which is signaled continuously. PMD_SIGNAL. indication is intended to be an indicator of the presence of the optical signal.

The value of the SIGNAL_DETECT parameter shall be generated according to the conditions defined in Table 75.4 for 10GBASE–PR and 10/1GBASE–PRX type PMDs. The ONU PMD receiver is not required to verify whether a compliant 10GBASE–PR signal is being received.

75.3.5.2 OLT PMD Signal Detect The response time for the PMD Signal Detect function for the burst mode upstream signal may be longer or shorter than a burst length; thus, it may not fulfill the traditional requirements placed on Signal Detect. PMD_SIGNAL.indication is intended to be an indicator of optical signal presence. The signal detect function in the OLT may be realized in the PMD or the Clause 76 PMA sublayer.

The value of the SIGNAL_DETECT parameter shall be generated according to the conditions defined in Table 75.4 for PMDs defined in Clause 75. The 10GBASE–PR–D PMD receiver is not required to verify whether a compliant

TABLE 75.4 SIGNAL_DETECT Value Definitions for Clause 75 PMDs

Receive conditions			Signal_ detect Value
10GBASE–PR–D1, 10GBASE–PR–D2, 10GBASE–PR–D3	10GBASE–PR–U1, 10GBASE–PR–U3, 10/1GBASE– PRX–U1, 10/1GBASE– PRX–U2, 10/1GBASE–PRX–U3	10/1GBASE– PRX–D1, 10/1GBASE– PRX–D2, 10/1GBASE– PRX–D3	
Average input optical power ≤ Signal Detect Threshold (min) in Table 75.6 at the specified receiver wavelength	Average input optical power ≤ Signal Detect Threshold (min) in Table 75.11 at the specified receiver wavelength	Average input optical power ≤ Signal Detect Threshold (min) in Table 75.7 at the specified receiver wavelength	FAIL
Average input optical power ≥ Receive sensitivity (max) in Table 75.6 with a compliant 10GBASE–PR signal input at the specified receiver wavelength	Average input optical power ≥ Receive sensitivity (max) in Table 75.11 with a compliant 10GBASE–PR signal input at the specified receiver wavelength	Average input optical power ≥ Receive sensitivity (max) in Table 75.7 with a compliant 1000GBASE–PX signal input at the specified receiver wavelength	OK
All other conditions	All other conditions	All other conditions	Unspecified

10GBASE–PR signal is being received. Similarly, the 10/1GBASE–PRX–D PMD receiver is not required to verify whether a compliant 1000BASE–PX signal is being received.

75.3.5.3 10GBASE–PR and 10/1GBASE–PRX Signal Detect Functions The Signal Detect value definitions for Clause 75 PMDs are shown in Table 75.4.

75.3.6 PMD Transmit Enable Function for ONU

PMD_SIGNAL. request (tx_enable) is defined for all ONU PMDs specified in Clause 75.

PMD_SIGNAL. request (tx_enable) is asserted prior to data transmission by the ONU PMDs.

75.4 PMD TO MDI OPTICAL SPECIFICATIONS FOR 10/10G–EPON AND 10/1G–EPON OLT PMDS

This subclause details the PMD to MDI optical specifications for 10/10G–EPON and 10/1G–EPON OLT

PMDs, as specified in 75.2. Specifically, 75.4.1 defines the OLT transmit parameters, while 75.4.2 defines the OLT receive parameters.

The operating ranges for PR and PRX power budget classes are defined in Table 75.1. A PR or PRX compliant transceiver operates over the media types listed in Table 75.14 according to the specifications described in 75.9. A transceiver which exceeds the operational range requirement while meeting all other optical specifications is considered compliant.

NOTE—The specifications for OMA have been derived from extinction ratio and average launch power (minimum) or receiver sensitivity (maximum). The calculation is defined in 58.7.6.

75.4.1 Transmitter Optical Specifications

The signaling speed, operating wavelength, side mode suppression ratio, average launch power, extinction ratio, return loss tolerance, OMA, eye, and Transmitter and Dispersion Penalty (TDP) for transmitters making part of the 10/10G–EPON and 10/1G–EPON OLT PMDs (as specified in 75.2) shall meet the specifications defined in Table 75.5 per measurement techniques described in 75.7. Their RIN15OMA should meet the value listed in Table 75.5 per measurement techniques described in 75.7.8. Note that there are only two groups of transmit parameters. The first group is shared by 10GBASE–PR–D1, 10/1GBASE–PRX–D1, 10GBASE–PR–D3, and 10/1GBASE–PRX–D3. The second group is shared by 10GBASE–PR–D2 and 10/1GBASE–PRX–D2.

The relationship between OMA, extinction ratio, and average power is described in 58.7.6 and illustrated in Figure 75.4 for a compliant transmitter. Note that the OMAmin and AVPmin are calculated for ER = 9 dB, where AVPmin represents the Average launch power (min) as presented in Table 75.5. The transmitter specifications are further relaxed by allowing lower ER = 6 dB while maintaining the OMAmin and AVPmin constant. The shaded area indicates a compliant part.

75.4.2 Receiver Optical Specifications

The signaling speed, operating wavelength, overload, stressed sensitivity, reflectivity, and signal detect for receivers forming part of the 10/10G–EPON and 10/1G–EPON OLT PMDs (as specified in 75.2) shall meet the specifications

TABLE 75.5   PR and PRX Type OLT PMD Transmit Characteristics

Description	10GBASE–PR–D1, 10GBASE–PR–D3, 10/1GBASE–PRX–D1, 10/1GBASE–PRX–D3	10GBASE–PR–D2, 10/1GBASE–PRX–D2	Unit
Signaling speed (range)	10.3125 ± 100 ppm	10.3125 ± 100 ppm	GBd
Wavelength (range)	1575 to 1580	1575 to 1580	nm
Side mode suppression ratio (min)[a]	30	30	dB
Average launch power (max)	5	9	dBm
Average launch power (min)[b]	2	5	dBm
Average launch power of OFF transmitter (max)	−39	−39	dBm
Extinction ratio (min)	6	6	dB
$RIN_{15}OMA$ (max)	−128	−128	dB/Hz
Launch OMA (min)[b]	3.91 (2.46)	6.91 (4.91)	dBm (mW)
Transmitter eye mask definition (X1, X2, X3, Y1, Y2, Y3)[c]	(0.25, 0.40, 0.45, 0.25, 0.28, 0.40)	(0.25, 0.40, 0.45, 0.25, 0.28, 0.40)	UI
Optical return loss tolerance (max)	15	15	dB
Transmitter reflectance (max)	−10	−10	dB
Transmitter and dispersion penalty (max)	1.5	1.5	dB
Decision timing offset for transmitter and dispersion penalty	±0.05	±0.05	UI

[a] Transmitter is a single longitudinal mode device. Chirp is allowed such that the total optical path penalty does not exceed that found in Table 75B.2.
[b] Minimum average launch power and minimum launch OMA are valid for ER = 9 dB (see Figure 75.4 for details).
[c] As defined in Figure 75.8.

defined in Table 75.6 for 10/10G–EPON OLT PMDs and in Table 75.7 for 10/1G–EPON OLT PMDs, per measurement techniques defined in 75.7. Their unstressed receive characteristics should meet the values listed in Table 75.6 and Table 75.7 per measurement techniques described in 75.7.11.

Either the damage threshold included in Table 75.6 or Table 75.7 shall be met, or the receiver shall be labeled to indicate the maximum optical input power level to which it can be continuously exposed without damage.

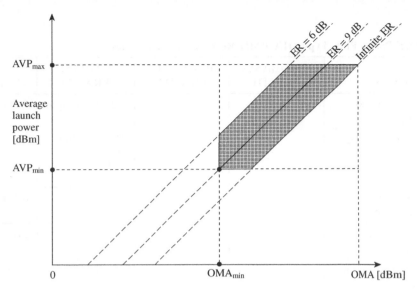

Figure 75.4 Graphical representation of region of PR–D type transmitter compliance.

TABLE 75.6 PR type OLT PMD Receive Characteristics

Description	10GBASE–PR–D1	10GBASE–PR–D2, 10GBASE–PR–D3	Unit
Signaling speed (range)	10.3125 ± 100 ppm	10.3125 ± 100 ppm	GBd
Wavelength (range)	1260 to 1280	1260 to 1280	nm
Bit error ratio (max)[a]	10^{-3}	10^{-3}	–
Average receive power (max)	−1	−6	dBm
Damage threshold (max)[b]	0	−5	dBm
Receiver sensitivity (max)	−24	−28	dBm
Receiver sensitivity OMA (max)	−23.22 (4.77)	−27.22 (1.90)	dBm (μW)
Signal detect threshold (min)	−45	−45	dBm
Receiver reflectance (max)	−12	−12	dB
Stressed receive sensitivity (max)[c]	−21	−25	dBm
Stressed receive sensitivity OMA (max)	−20.22 (9.51)	−24.22 (3.79)	dBm (μW)
Vertical eye–closure penalty[d]	2.99	2.99	dB
$T_{receiver\_settling}$ (max)[e]	800	800	ns
Stressed eye jitter	0.3	0.3	UI pk to pk
Jitter corner frequency for a sinusoidal jitter	4	4	MHz
Sinusoidal jitter limits for stressed receiver conformance test (min, max)	(0.05, 0.15)	(0.05, 0.15)	UI

[a] The BER of 10^{-12} is achieved by the utilization of FEC as described in 76.3.
[b] Direct ONU–OLT connection may result in damage of the receiver.
[c] The stressed receiver sensitivity is mandatory.
[d] Vertical eye closure penalty and the jitter specifications are test conditions for measuring stressed receiver sensitivity. They are not required characteristics of the receiver.
[e] $T_{receiver\_settling}$ represents an upper bound. Optics with better performance may be used in compliant implementations, since the OLT notifies the ONUs of its requirements in terms of the $T_{receiver\_settling}$ time via the SYNCTIME parameter (see 77.3.3.2).

TABLE 75.7 PRX Type OLT PMD Receive Characteristics

Description	10/1GBASE–PRX–D1	10/1GBASE–PRX–D2	10/1GBASE–PRX–D3	Unit
Signaling speed (range)			1.25 ± 100 ppm	GBd
Wavelength (range)			1260 to 1360	nm
Bit error ratio (max)			10^{-12}	
Average receive power (max)			−9.38	dBm
Damage threshold (max)			−5.00	dBm
Receiver sensitivity (max)			−29.78	dBm
Receiver sensitivity OMA (max)			−29.00 (1.26)	dBm (μW)
Signal detect threshold (min)	same as 1000BASE–PX10–D receive parameters (see Table 60.5)	same as 1000BASE–PX20–D receive parameters (see Table 60.8)	−45	dBm
Receiver reflectance (max)			−12	dB
Stressed receive sensitivity (max)			−28.38[a]	dBm
Stressed receive sensitivity OMA (max)			−27.60 (1.74)	dBm (μW)
Vertical eye–closure penalty[b]			1.4	dB
$T_{receiver\_settling}$ (max)[c]			400	ns
Stressed eye jitter			0.28	UI pk to pk
Jitter corner frequency for a sinusoidal jitter			637	kHz
Sinusoidal jitter limits for stressed receiver conformance test (min, max)			(0.05, 0.15)	UI

[a] The stressed receiver sensitivity is mandatory.

[b] Vertical eye closure penalty and the jitter specifications are test conditions for measuring stressed receiver sensitivity. They are not required characteristics of the receiver.

[c] $T_{receiver\_settling}$ represents an upper bound. Optics with better performance may be used in compliant implementations, since the OLT notifies the ONUs of its requirements in terms of the $T_{receiver\_settling}$ time via the SYNCTIME parameters (see 77.3.3.2).

The damage threshold included in Tables 75.6 and 75.7 does not guarantee direct ONU–OLT connection, which may result in damage of the receiver. If direct ONU–OLT connection is necessary, optical attenuators and/or equivalent loss components should be inserted to decrease receive power below the damage threshold.

75.5 PMD TO MDI OPTICAL SPECIFICATIONS FOR 10/10G–EPON AND 10/1G–EPON ONU PMDS

This subclause details the PMD to MDI optical specifications for 10/10G–EPON and 10/1G–EPON ONU PMDs, as specified in 75.2. Specifically, 75.5.1 defines the ONU transmit parameters, while 75.5.2 defines the ONU receive parameters.

The operating ranges for PR and PRX power budget classes are defined in Table 75.1. A PR or PRX compliant transceiver operates over the media types listed in Table 75.14 according to the specifications described in 75.9. A transceiver which exceeds the operational range requirement while meeting all other optical specifications is considered compliant.

NOTE—The specifications for OMA have been derived from extinction ratio and average launch power (minimum) or receiver sensitivity (maximum). The calculation is defined in 58.7.6.

75.5.1 Transmitter Optical Specifications

The signaling speed, operating wavelength, spectral width (for 10/1G–EPON ONU PMDs) or side mode suppression ratio (for 10/10G–EPON ONU PMDs), average launch power, extinction ratio, return loss tolerance, OMA, eye and TDP for transmitters forming part of the 10/10G–EPON and 10/1G–EPON ONU PMDs (as specified in 75.2) shall meet the specifications defined in Table 75.8 for 10/10G–EPON ONU PMDs and in Table 75.9 for 10/1G–EPON ONU PMDs, per measurement techniques described in 75.7.

Their RIN15OMA should meet the value listed in Table 75.8 or Table 75.9, as appropriate, per measurement techniques described in 75.7.8.

The relationship between OMA, extinction ratio and average power is described in 58.7.6 and illustrated in Figure 75.5 for a compliant transmitter. Note that the OMAmin and AVPmin are calculated for ER = 6 dB.

The transmitter average launch power specifications are further relaxed by allowing ER higher than 6 dB while maintaining the OMAmin constant. The shaded area indicates a compliant part.

75.5.2 Receiver Optical Specifications

The signaling speed, operating wavelength, overload, stressed sensitivity, reflectivity, and signal detect for receivers forming part of the 10/10G–EPON

TABLE 75.8   PR type ONU PMD Transmit Characteristics

Description	10GBASE-PR-U1	10GBASE-PR-U3	Unit
Signaling speed (range)	10.3125 ± 100 ppm	$10.3125 \pm 100V$ ppm	GBd
Wavelength (range)	1260 to 1280	1260 to 1280	nm
Side Mode Suppression Ratio (min)[a]	30	30	dB
Average launch power (max)	4	9	dBm
Average launch power (min)[b]	−1	4	dBm
Average launch power of OFF transmitter (max)	−45	−45	dBm
Extinction ratio (min)	6	6	dB
$RIN_{15}OMA$ (max)	−128	−128	dB/Hz
Launch OMA (min)[b]	−0.22 (0.95)	4.78 (3.01)	dBm (mW)
Transmitter eye mask definition (X1, X2, X3, Y1, Y2, Y3)[c]	(0.25, 0.40, 0.45, 0.25, 0.28, 0.40)	(0.25, 0.40. 0.45, 0.25, 0.28, 0.40)	UI
T_{on} (max)	512	512	ns
T_{off} (max)	512	512	ns
Optical return loss tolerance (max)	15	15	dB
Transmitter reflectance (max)	−10	−10	dB
Transmitter and dispersion penalty (max)[d]	3.0	3.0	dB
Decision timing offset for transmitter and dispersion penalty	±0.0.625	±0.0.625	UI

[a] Transmitter is a single longitudinal mode device. Chirp is allowed such that the total optical path penalty does not exceed that found in Table 75B.2.
[b] Minimum average launch power and minimum launch OMA are valid for ER = 6 dB (see Figure 75.5 for details).
[c] As defined in Figure 75.8.
[d] If a transmitter has a lower TDP, the minimum transmitter launch OMA (OMA_{min}) and average minimum launch power (AVP_{min}) may be relaxed by the amount 3.0 dB—TDP.

ONU and 10/1G–EPON ONU PMDs (as specified in 75.2) shall meet the specifications defined in Table 75.11 for Clause 75 ONU PMDs, per measurement techniques defined in 75.7. Their unstressed receive characteristics should meet the values listed in Table 75.11 per measurement techniques described in 75.7.11. Either the damage threshold included in Table 75.11 shall be met, or the receiver shall be labeled to indicate the maximum optical input power level to which it can be continuously exposed without damage. (Figure 75.6)

TABLE 75.9 PRX Type ONU PMD Transmit Characteristics

Description	10/1GBASE-PRX-U1	10/1GBASE-PRX-U2	10/1GBASE-PRX-U3	Unit
Signaling speed (range)			1.25 ± 100 ppm	GBd
Wavelength[a] (range)			1260 to 1360	nm
RMS spectral width (max)			see[b]	nm
Average launch power (max)			5.62	dBm
Average launch power (min)[c]			0.62	dBm
Average launch power of OFF transmitter (max)			−45	dBm
Extinction ratio (min)			6	dB
$RIN_{15}OMA$ (max)	same as 1000BASE–PX10–U transmit parameters (see Table 60.3)	same as 1000BASE–PX20–U transmit parameters (see Table 60.6)	−115	dB/Hz
Launch OMA (min)[c]			1.40 (1.38)	dBm (mW)
transmitter eye mask definition (X1, X2, Y1, Y2, Y3)[d]			(0.22, 0.375, 0.20, 0.20, 0.30)	UI
T_{on} (max)			512	ns
T_{off} (max)			512	ns
Optical return loss tolerance (max)			15	dB
Transmitter reflectance (max)			−10	dB
Transmitter and dispersion penalty (max)			1.4	dB
Decision timing offset for transmitter and dispersion penalty			±0.125	UI

[a] This represents the range of center wavelength ± 1σ of the rms spectral width.
[b] If the transmitter employs a Fabry–Perot laser, the RMS spectral width shall comply with Table 75.10. If the transmitter employs a DFB laser, the side mode suppression ratio (min) shall be 30 dB.
[c] Minimum average launch power and minimum launch OMA are valid for ER = 6 dB.
[d] As defined in Figure 75.7.

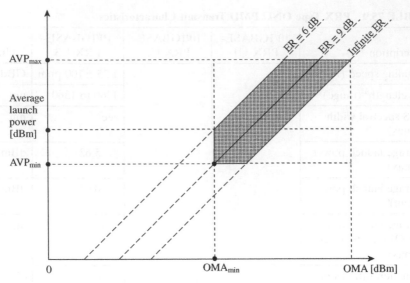

Figure 75.5 Graphical representation of region of PR–U type transmitter compliance.

TABLE 75.10   10/1GBASE-PRX–U3 Transmitter Spectral Limits

Center Wavelength	RMS Spectral Width (max)[a]	RMS Spectral Width to Achieve Epsilon $\varepsilon \leq 0.08$ (Informative)
nm	nm	nm
1260	0.59	0.5
1270	0.7	0.59
1280	0.87	0.74
1290	1.14	0.97
1300	1.64	1.39
1304	1.98	1.67
1305	2.09	1.77
1308	2.4	2
1317	2.4	2
1320	2.07	1.75
1321	1.98	1.67
1330	1.4	1.18
1340	1.06	0.89
1350	0.86	0.72
1360	0.72	0.61

[a] These limits for the 10/1GBASE-PRX-U3 transmitter are illustrated in Figure 75.6. The equation used to calculate these values is detailed in 60.7.2. Limits at intermediate wavelengths may be found by interpolation.

TABLE 75.11 PR and PRX Type ONU PMD Receive Characteristics

Description	10GBASE-PR-U1, 10/1GBASE-PRX-U1, 10/1GBASE-PRX-U2	10GBASE-PR-U3, 10/1GBASE-PRX-U3	Unit
Signaling speed (range)	10.3125 ± 100 ppm	10.3125 ± 100 ppm	GBd
Wavelength (range)	1575 to 1580	1575 to 1580	nm
Bit error ratio (max)[a]	10^{-3}	10^{-3}	—
Average receive power (max)	0	−10	dBm
Damage threshold (max)[b]	1	−9	dBm
Receiver sensitivity (max)	−20.50	−28.50	dBm
Receiver sensitivity OMA (max)	−18.59 (13.84)	−26.59 (2.19)	dBm (μW)
Signal detect threshold (min)	−44	−44	dBm
Receiver reflectance (max)	−12	−12	dB
Stressed receive sensitivity (max)[c]	−19	−27	dBm
Stressed receive sensitivity OMA (max)	−17.09 (19.55)	−25.09(3.10)	dBm (μW)
Vertical eye–closure penalty[d]	1.5	1.5	dB
Stressed eye jitter (min)	0.3	0.3	UI pk to pk
Jitter corner frequency for a sinusoidal jitter	4	4	MHz
Sinusoidal jitter limits for stressed receiver conformance test (min, max)	(0.05, 0.15)	(0.05, 0.15)	UI

[a] The BER of 10^{-12} is achieved by the utilization of FEC as described in 76.3.
[b] Direct ONU-OLT connection may result in damage of the receiver.
[c] The stressed receiver sensitivity is mandatory over the entire PR-D transmitter compliance region, as illustrated in Figure 75.4.
[d] Vertical eye closure penalty and the jitter specifications are test conditions for measuring stressed receiver sensitivity. They are not required characteristics of the receiver.

The damage threshold included in Table 75.11 does not guarantee direct ONU–OLT connection, which may result in damage of the receiver. If direct ONU–OLT connection is necessary, optical attenuators and/or equivalent loss components should be inserted to decrease receive power below damage threshold.

75.6 DUAL-RATE (COEXISTENCE) MODE

To support coexistence of 10G–EPON and 1G–EPON ONUs on the same outside plant, the OLT may be configured to use a dual-rate mode. Dual-rate mode supports transmission and/or reception of both 10 Gb/s and 1 Gb/s data

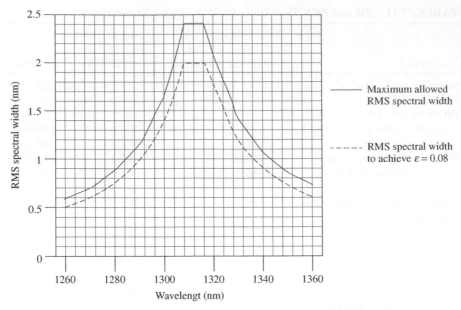

Figure 75.6   10/1GBASE–PRX–U3 transmitter spectral limits.

TABLE 75.12   PMD Coexistence Mapping for Dual-rate Mode Option[a]

Direction of Dual-rate Operation	OLT PMD Combination	ONU PMDs Coexisting on the Same ODN
Downstream	1000BASE-PX-D 10/1GBASE-PRX-D	(1) 1000BASE-PX-U (2) 10/1GBASE-PRX-U
Upstream	10GBASE-PX-D 10/1GBASE-PRX-D	(1) 10GBASE-PR-U (2) 10/1GBASE-PRX-U
Downstream and upstream	1000BASE-PX-D 10GBASE-PR-D	(1) 1000BASE-PX-U (2) 10/1GBASE-PRX-U (3) 10GBASE-PR-U

[a] Only PMDs with compatible power budgets can be connected to the same ODN.

rates, and can be introduced as options for 10G–EPON OLTs, functionally combining PMDs supporting 10 Gb/s and 1 Gb/s data rates.

Table 75.12 depicts PMD coexistence mapping for dual-rate mode options.

75.6.1   Downstream Dual-Rate Operation

When the downstream dual-rate operation is enabled, the OLT transmits both 10 Gb/s and 1 Gb/s downstream signals in a WDM manner. The OLT should meet both 10 Gb/s and 1 Gb/s specifications defined in Table 75.5 (10GBASE–PR–D transmit characteristics) and in Table 60.3 or Table 60.6 (1000BASE–PX–D transmit characteristics).

75.6.2 Upstream Dual-Rate Operation

When the upstream dual-rate operation is enabled, the OLT receives both 10 Gb/s and 1 Gb/s upstream signals in a TDMA manner. Further implementation details are described in Annex 75A. The OLT should meet both 10 Gb/s and 1 Gb/s specifications defined in Table 75.6 (10GBASE–PR–D receive characteristics), and in Tables 60.5 and 60.8 (1000BASE–PX–D receive characteristics), and Table 75.7 (10/1GBASE–PRX–D receive characteristics).

NOTE—The damage threshold values in Table 60.5, Table 60.8, and Table 75.7 are considerably higher than those in Table 75.6; the dual-rate PMD should be labeled appropriately.

75.7 DEFINITIONS OF OPTICAL PARAMETERS AND MEASUREMENT METHODS

When measuring jitter at TP1 and TP5, it is recommended that jitter contributions at frequencies below receiver corner frequencies (i.e., 4 MHz for 10.3125 GBd receiver and 637 kHz for 1.25 GBd receiver) are filtered at the measurement unit. The following subclauses describe definitive patterns and test procedures for certain PMDs of this standard. Implementers using alternative verification methods should ensure adequate correlation and allow adequate margin such that specifications are met by reference to the definitive methods. All optical measurements, except TDP and RIN15OMA shall be made through a short patch cable between 2 m and 5 m in length.

75.7.1 Insertion Loss

Insertion loss for SMF fiber optic cabling (channel) is defined at 1270 nm, 1310 nm, or 1577 nm, depending on the particular PMD. A suitable test method is described in ITU–T G.650.1.

75.7.2 Allocation for Penalties in 10G–EPON PMDs

All the receiver types specified in Clause 75 are required to tolerate a path penalty not exceeding 1 dB to account for total degradations due to reflections, intersymbol interference, mode partition noise, laser chirp and detuning of the central wavelength, including chromatic dispersion penalty. All the transmitter types specified in Clause 75 introduce less than 1 dB of optical path penalty over the channel. The path penalty is a component of transmitter and

TABLE 75.13   Test Patterns

Test	10 Gb/s pattern[a]	1 Gb/s pattern	Related subclause
Average optical power	1 or 3	Valid 8B/10B	75.7.5
OMA (modulated optical power)	Square	Idles	75.7.7
Extinction ratio	1 or 3	Idles	75.7.6
Transmit eye	1 or 3	Valid 8B/10B	75.7.7
RIN$_{15}$OMA	Square	Idles	75.7.8
Wavelength, spectral width	1 or 3	Valid 8B/10B	75.7.4
Side mode suppression ratio	1 or 3	Valid 8B/10B	–
VECP calibration	2 or 3	Jitter frame	75.7.12
Receiver sensitivity	1 or 3	Random frame	75.7.11
Receiver overload	1 or 3	Valid 8B/10B	–
Stressed receive sensitivity	2 or 3	Random frame	75.7.12
Transmitter and dispersion penalty	2 or 3	Random frame	75.7.10
Jitter	2 or 3	Jitter frame	75.7.13
Laser On/Off	1 or 3	Valid 8B/10B	75.7.14
Receiver settling	1 or 3	Valid 8B/10B	75.7.15

[a] Individual 10 Gb/s test patterns are described in 52.9.1.2 for a square wave and 52.9.1.1 for test patterns represented by numbers.

dispersion penalty (TDP), which is specified in Table 75.5, Tables 75.8 and 75.9 and described in 58.7.9.

75.7.3   Test Patterns

Two types of test patterns are used for testing of 10 Gb/s optical PMDs: square wave (52.9.1.2) and patterns 1, 2, or 3 (52.9.1.1). These 10 Gb/s test patterns for 10GBASE–PR and 10/1GBASE–PRX are in Table 75.13. Two types of test frames (random and jitter [59.7.1]) are used for 1 Gb/s tests relevant to the 10/1GBASE–PRX PHY. All test patterns are listed in Table 75.13.

75.7.4   Wavelength and Spectral Width Measurement

The center wavelength and spectral width (RMS) shall meet the specifications when measured according to TIA-455-127-A under modulated conditions using an appropriate PRBS or a valid 10GBASE–PR signal, 1000BASE–X signal, or another representative test pattern.

NOTE—The allowable range of central wavelengths is narrower than the operating wavelength range by the actual RMS spectral width at each extreme.

75.7.5   Optical Power Measurements

Optical power shall meet specifications according to the methods specified in ANSI/EIA-455-95. A measurement may be made with the port transmitting any valid encoded 8B/10B or 64B/66B data stream.

75.7.6   Extinction Ratio Measurements

The extinction ratio shall meet the specifications when measured according to IEC 61820-2-2 with the port transmitting a repeating idle pattern /I2/ ordered_set (see 36.2.4.12) or valid 10GBASE–PR signal, and with minimal back reflections into the transmitter, lower than −20 dB. The test receiver has the frequency response as specified for the transmitter optical waveform measurement.

75.7.7   Optical Modulation Amplitude (OMA) Test Procedure

A description of OMA measurements for 1 Gb/s PHYs is found in 58.7.5. The OMA measurements for 10 Gb/s PHYs shall be compliant with the description found in 52.9.5.

75.7.8   Relative Intensity Noise Optical Modulation Amplitude (RINxOMA) Measuring Procedure

This procedure describes a component test that may not be appropriate for a system level test depending on the implementation. If used, the procedure shall be performed as described in 52.9.6 for 10 Gb/s PHYs and in 58.7.7 for 1 Gb/s PHYs.

75.7.9   Transmit Optical Waveform (Transmit Eye)

The required transmitter pulse shape characteristics are specified in the form of a mask of the transmitter eye diagram as shown in Figure 75.7 for

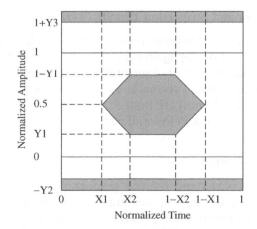

Figure 75.7   Transmitter eye mask definition for upstream direction of 10/1GBASE–PRX PMD.

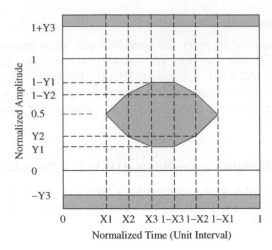

Figure 75.8 Transmitter eye mask definition for downstream direction of 10/1GBASE–PRX PMD and both directions of 10GBASE–PR PMD.

upstream direction of 10/1GBASE–PRX PMD and Figure 75.8 for downstream direction of 10/1GBASE–PRX PMD and both directions of 10GBASE–PR PMD.

The measurement procedure is described in 58.7.8 for 1 Gb/s PHYs and 52.9.7 for 10 Gb/s PHYs and references therein. The eye shall comply to the mask of the eye using a fourth-order Bessel-Thomson receiver response as defined in 60.7.8 for 1 Gb/s PMD transmitters and 52.9.7 for 10 Gb/s PMD transmitters.

75.7.10 Transmitter and Dispersion Penalty (TDP)

TDP measurement tests transmitter impairments, including chromatic dispersion effects, due to signal propagation in SMF used in PON. Possible causes of impairment include intersymbol interference, jitter, and RIN. Meeting the separate requirements (e.g., eye mask, spectral characteristics) does not in itself guarantee the TDP. The TDP limit shall be met. See 58.7.9 for details of the measurement for 1 Gb/s PHYs and 52.9.10 for 10 Gb/s PHYs.

75.7.11 Receive Sensitivity

Receiver sensitivity is defined for the random pattern test frame, or test pattern 1, or test pattern 3, and an ideal input signal quality with the specified extinction ratio. The measurement procedure is described in 58.7.10 for 1 Gb/s PHYs and 52.9.8 for 10 Gb/s PHYs. The sensitivity shall be met for the bit error ratio defined in Table 75.6, Table 75.7, or Table 75.11 as appropriate.

75.7.12 Stressed Receiver Conformance Test

Compliance with stressed receiver sensitivity is mandatory for the following PMDs: 10GBASE–PR–D1, 10GBASE–PR–D2, 10GBASE–PR–D3, 10GBASE–PR–U1, 10GBASE–PR–U3, 10/1GBASE–PRX–D3, 10/1GBASE–PRX–U1, 10/1GBASE–PRX–U2, and 10/1GBASE–PRX–U3. The stressed receiver conformance test is intended to screen against receivers with poor frequency response or timing characteristics that could cause errors when combined with a distorted but compliant signal. To be compliant with stressed receiver sensitivity, the receiver shall meet the specified bit error ratio at the power level and signal quality defined in Table 75.6, Table 75.7, or Table 75.11 as appropriate, according to the measurement procedures of 58.7.11 for 1 Gb/s PHYs and 52.9.9 for 10 Gb/s PHYs.

75.7.13 Jitter Measurements

Jitter measurements for 1 Gb/s are described in 58.7.12. Jitter measurements for 10 Gb/s are described in 52.8.1.

75.7.14 Laser on/off Timing Measurement

The laser on/off timing measurement procedure is described in 60.7.13.1 with the following changes:

  a. T_{on} is defined in 60.7.13.1.1, and its value is less than 512 ns (defined in Tables 75.8 and 75.9).
  b. $T_{receiver\_settling}$ is defined in 60.7.13.2.1, and its value is defined in Tables 75.6 and 75.7.
  c. T_{CDR} is defined in 76.4.2.1, and its value is less than 400 ns.
  d. $T_{code\_group\_align}$ is defined in 36.6.2.4, and its value is less than 4 ten bit code–groups for 1 Gb/s PHYs, and is defined as 0 for 10 Gb/s PHYs.
  e. T_{off} is defined in 60.7.13.11.1, and its value is less than 512 ns (defined in Tables 75.8 and 75.9).

75.7.15 Receiver Settling Timing Measurement

75.7.15.1 Definitions Denote $T_{receiver\_settling}$ as the time beginning from the time that the optical power in the receiver at TP7 reaches the conditions specified in 75.7.12 and ending at the time that the electrical signal after the PMD at TP8 reaches within 15% of its steady state parameter (average power, jitter) (see Table 75.6 for 10GBASE–PR–D1, 10GBASE–PR–D2, and 10GBASE–PR–D3, and Table 75.7 for 10/1GBASE–PRX–D1, 10/1GBASE–PRX–D2, and 10/1GBASE–PRX–D3). Treceiver_settling is presented in Figure 75.9. The data transmitted may be any valid 64B/66B symbols (or a specific power synchronization sequence). The optical signal at TP7, at the beginning of the

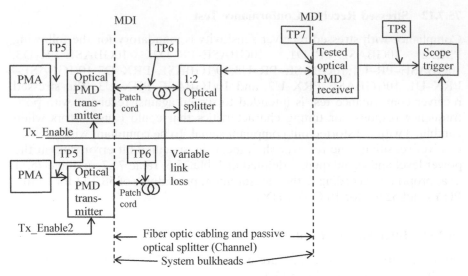

Figure 75.9 Receiver settling time measurement setup.

locking, may have any valid 64B/66B pattern, optical power level, jitter, or frequency shift matching the standard specifications.

75.7.15.2 Test Specification Figure 75.9 illustrates the test setup for the OLT PMD receiver (upstream) Treceiver_settling time. The optical PMD transmitter has well–known parameters, with a fixed known Ton time. After Ton time the parameters of the reference transmitter, at TP6 and therefore at TP7, reach within 15% of its steady state values as specified in Table 75.8 for 10GBASE–PR–U1 and 10GBASE–PR–U3 and Table 75.9 for 10/1GBASE–PRX–U1, 10/1GBASE–PRX–U2 and 10/1GBASE–PRX–U3.

Define $T_{receiver\_settling}$ time as the time from the Tx_Enable assertion, minus the known T_{on} time, to the time the electrical signal at TP8 reaches within 15% of its steady state conditions.

Conformance should be assured for an optical signal at TP7 with any level of its specified parameters before the Tx_Enable assertion. Especially the Treceiver_settling time must be met in the following scenarios:

- Switching from a "weak" (minimal received power at TP7) ONU to a "strong" (maximal received power at TP7) ONU, with minimal guard band between.
- Switching from a "strong" ONU to a "weak" ONU, with minimal guard band between.
- Switching from noise level, with maximal duration interval, to "strong" ONU power level.

A non-rigorous way to describe this test setup would be (using a transmitter with a known T_{on}).

For a tested PMD receiver with a declared $T_{receiver\_settling}$ time, measure all PMD receiver electrical parameters at TP8 after $T_{receiver\_settling}$ from the TX_ENABLE trigger minus the reference transmitter T_{on}, reassuring conformance to within 15% of its specified steady state values.

75.8   ENVIRONMENTAL, SAFETY, AND LABELING

75.8.1   General Safety

All equipment subject to this clause shall conform to IEC 60950–1.

75.8.2   Laser Safety

10GBASE–PR and 10/1GBASE–PRX optical transceivers shall conform to Class 1 laser requirements as defined in IEC 60825–1 and IEC 60825–2, under any condition of operation. This includes single fault conditions whether coupled into a fiber or out of an open bore.

Conformance to additional laser safety standards may be required for operation within specific geographic regions.

Laser safety standards and regulations require that the manufacturer of a laser product provide information about the product's laser, safety features, labeling, use, maintenance, and service. This documentation explicitly defines requirements and usage restrictions on the host system necessary to meet these safety certifications.

75.8.3   Installation

It is recommended that proper installation practices, as defined by applicable local codes and regulation, be followed in every instance in which such practices are applicable.

75.8.4   Environment

The 10GBASE–PR and 10/1GBASE–PRX operating environment specifications are as defined in 52.11, as defined in 52.11.1 for electromagnetic emission, and as defined in 52.11.2 for temperature, humidity, and handling.

See Annex 67A for additional environmental information. Two optional temperature ranges are defined in Table 60.13. Implementations shall be declared as compliant over one or both complete ranges, or not so declared (compliant over parts of these ranges or another temperature range).

75.8.5 PMD Labeling

The 10GBASE–PR and 10/1GBASE–PRX labeling recommendations and requirements are as defined in 52.12.

Defined PMDs are as follows:

- 10/1GBASE–PRX–D1
- 10/1GBASE–PRX–D2
- 10/1GBASE–PRX–D3
- 10GBASE–PR–D1
- 10GBASE–PR–D2
- 10GBASE–PR–D3
- 10/1GBASE–PRX–U1
- 10/1GBASE–PRX–U2
- 10/1GBASE–PRX–U3
- 10GBASE–PR–U1
- 10GBASE–PR–U3

Each field-pluggable component shall be clearly labeled with its operating temperature range over which compliance is ensured.

75.9 CHARACTERISTICS OF THE FIBER OPTIC CABLING

The 10GBASE–PR and 10/1GBASE–PRX fiber optic cabling shall meet the dispersion specifications defined in IEC 60793–2 and ITU–T G.652, as shown in Table 75.14. The fiber optic cabling consists of one or more sections of fiber optic cable and any intermediate connections required to connect sections together.

It also includes a connector plug at each end to connect to the MDI. The fiber optic cabling spans from one MDI to another MDI, as shown in Figure 75.3.

75.9.1 Fiber Optic Cabling Model

The fiber optic cabling model is shown in Figure 75.3.

NOTE—The optical splitter presented in Figure 75.3 may be replaced by a number of smaller 1:n splitters such that a different topology may be implemented while preserving the link characteristics and power budget as defined in Tables 75B–1 and 75B–2.

The maximum channel insertion losses shall meet the requirements specified in Table 75.1. Insertion loss measurements of installed fiber cables are made in accordance with IEC 61280–4–2:2000. The fiber optic cabling model

TABLE 75.14 Optical Fiber and Cable Characteristics

Description[a]	IEC 60793-2 B1.1, B1.3 SMF, ITU-T G.652, G.657 SMF[b]				Unit
Nominal wavelength[c]	1270	1310	1550	1577	nm
Cable attenuation (max)[d]	0.44	0.4	0.35	0.35	dB/km
Zero dispersion wavelength[e]	$1300 \le \lambda_0 \le 1324$				nm
Dispersion slope (max)	0.093				$ps/nm^2 \cdot km$

[a] The fiber dispersion values are normative, all other values in the table are informative.
[b] Other fiber types are acceptable if the resulting ODN meets channel insertion loss and dispersion requirements.
[c] Wavelength specified is the nominal wavelength and typical measurement wavelength. Power penalties at other wavelengths are accounted for.
[d] Attenuation for single-mode optical fiber cables for 1310 nm and 1550 nm is defined in ITU-T G.652. The attenuation values in the 1270 nm and 1577 nm windows were calculated using spectral attenuation modeling method (5.4.4) included in G.650.1 (06/2004) and the matrix coefficients included in Appendix III therein. 1310 nm (0.4 dB/km), 1380 nm (0.5 dB/km) and 1550 nm (0.35 dB/km) attenuation values were used as the input for the predictor model.
[e] See IEC 60793 or ITU-T G.652.

(channel) defined here is the same as a simplex fiber optic link segment. The term *channel* is used here for consistency with generic cabling standards.

75.9.2 Optical Fiber and Cable

The fiber optic cable requirements are satisfied by the fibers specified in IEC 60793–2 Type B1.1 (dispersion un–shifted SMF) and Type B1.3 (low water peak SMF), ITU–T G.652 and ITU–T G.657 (bend–insensitive SMF), as noted in Table 75.14.

75.9.3 Optical Fiber Connection

An optical fiber connection as shown in Figure 75.3 consists of a mated pair of optical connectors. The 10GBASE–PR or 10/1GBASE–PRX PMD is coupled to the fiber optic cabling through an optical connection and any optical splitters into the MDI optical receiver, as shown in Figure 75.3. The channel insertion loss includes the loss for connectors, splices and other passive components such as splitters, see Tables 75B.1 and 75B.2.

The channel insertion loss was calculated under the assumption of 14.5 dB loss for a 1:16 splitter/18.1 dB loss for a 1:32 splitter (ITU–T G.671 am 1). Unitary fiber attenuation for particular transmission wavelength is provided in Table 75.14. The number of splices/connectors is not predefined; the number of individual fiber sections between the OLT MDI and the ONU MDI is not defined. The only requirements are that the resulting channel insertion loss is within the limits specified in Table 75.1 and the maximum reach in Table 75.1 is not exceeded. Other fiber arrangements (e.g., increasing the split ratio while decreasing the fiber length) are supported as long as the limits for the channel insertion loss specified in Table 75.1 are observed.

The maximum discrete reflectance for single-mode connections shall be less than −26 dB.

75.9.4   Medium Dependent Interface (MDI)

The 10GBASE–PR or 10/1GBASE–PRX PMD is coupled to the fiber cabling at the MDI. The MDI is the interface between the PMD and the "fiber optic cabling" as shown in Figure 75.3. Examples of an MDI include the following:

a. Connectorized fiber pigtail
b. PMD receptacle

When the MDI is a remateable connection, it shall meet the interface performance specifications of IEC 61753–1. The MDI carries the signal in both directions for 10GBASE–PR or 10/1GBASE–PRX PMD and couples to a single fiber.

NOTE—Compliance testing is performed at TP2 and TP3 as defined in 75.3.2, not at the MDI.

75.10   PROTOCOL IMPLEMENTATION CONFORMANCE STATEMENT (PICS) PROFORMA FOR CLAUSE 75, PHYSICAL MEDIUM DEPENDENT (PMD) SUBLAYER AND MEDIUM FOR PASSIVE OPTICAL NETWORKS, TYPE 10GBASE–PR AND 10/1GBASE–PRX[1]

75.10.1   Introduction

The supplier of a protocol implementation that is claimed to conform to Clause 75, Physical Medium Dependent (PMD) sublayer and medium for passive optical networks, type 10GBASE–PR and 10/1GBASE–PRX, shall complete the following protocol implementation conformance statement (PICS) proforma.

A detailed description of the symbols used in the PICS proforma, along with instructions for completing the PICS proforma, can be found in Clause 21.

---

[1]Copyright release for PICS proformas: Users of this standard may freely reproduce the PICS proforma in this subclause so that it can be used for its intended purpose and may further publish the completed PICS.

75.10.2 Identification

75.10.2.1 Implementation Identification

Supplier[a]
Contact point for enquiries about the PICS[a]
Implementation Name(s) and Version(s)[a,c]
Other information necessary for full identification—for example, name(s) and version(s) for machines and/or operating systems; System Name(s)[b]

[a] Required for all implementations.
[b] May be completed as appropriate in meeting the requirements for the identification.
[c] The terms Name and Version should be interpreted appropriately to correspond with a supplier's terminology (e.g., Type, Series, Model).

75.10.2.2 Protocol Summary

Identification of protocol standard	IEEE Std 802.3av-2009, Clause 75, Physical Medium Dependent (PMD) sublayer and medium for passive optical networks, type 10GBASE-PR and 10/1GBASE-PRX
Identification of amendments and corrigenda to this PICS proforma that have been completed as part of this PICS	
Have any Exception items been required?	No [ ] Yes [ ]
(See Clause 21; the answer Yes means that the implementation does not conform to IEEE Std 802.3av-2009.)	
Date of Statement	

75.10.3 Major Capabilities/Options

Item	Feature	Subclause	Value/Comment	Status	Support
DTX	Transmit delay	75.3.1.1	Delay of 4 TQ (max) with variability 0.5 TQ (max)	M	Yes [ ]
DRX	Receive delay	75.3.1.1	Delay of 4 TQ (max) with variability 0.5 TQ (max)	M	Yes [ ]

(Continued)

Item	Feature	Subclause	Value/Comment	Status	Support
HT	High temperature operation	75.8.4	−5°C to 85°C	O	Yes [ ] No [ ]
LT	Low temperature operation	75.8.4	−40°C to 60°C	O	Yes [ ] No [ ]
*PR10U	10GBASE-PR-D1 or 10GBASE-PR-U1 PMD	75.4, 75.5	Maximum channel insertion loss of 20 dB	O/1	Yes [ ] No [ ]
*PR10D	10GBASE-PR-D1 or 10GBASE-PR-U1 PMD	75.4, 75.5	Maximum channel insertion loss of 20 dB	O/1	Yes [ ] No [ ]
*PR20U	10GBASE-PR-D2 or 10GBASE-PR-U1 PMD	75.4, 75.5	Maximum channel insertion loss of 24 dB	O/1	Yes [ ] No [ ]
*PR20D	10GBASE-PR-D2 or 10GBASE-PR-U1 PMD	75.4, 75.5	Maximum channel insertion loss of 24 dB	O/1	Yes [ ] No [ ]
*PR30U	10GBASE-PR-D3 or 10GBASE-PR-U3 PMD	75.4, 75.5	Maximum channel insertion loss of 29 dB	O/1	Yes [ ] No [ ]
*PR30D	10GBASE-PR-D3 or 10GBASE-PR-U3 PMD	75.4, 75.5	Maximum channel insertion loss of 29 dB	O/1	Yes [ ] No [ ]
*PRX10U	10/1GBASE-PRX-D1 or 10/1GBASE-PRX-U1 PMD	75.4, 75.5	Maximum channel insertion loss of 20 dB	O/1	Yes [ ] No [ ]
*PRX10D	10/1GBASE-PRX-D1 or 10/1GBASE-PRX-U1 PMD	75.4, 75.5	Maximum channel insertion loss of 20 dB	O/1	Yes [ ] No [ ]
*PRX20U	10/1GBASE-PRX-D2 or 10/1GBASE-PRX-U2 PMD	75.4, 75.5	Maximum channel insertion loss of 24 dB	O/1	Yes [ ] No [ ]
*PRX20D	10/1GBASE-PRX-D2 or 10/1GBASE-PRX-U2 PMD	75.4, 75.5	Maximum channel insertion loss of 24 dB	O/1	Yes [ ] No [ ]

Item	Feature	Subclause	Value/Comment	Status	Support
*PRX30U	10/1GBASE-PRX-D3 or 10/1GBASE-PRX-U3 PMD	75.4, 75.5	Maximum channel insertion loss of 29 dB	O/1	Yes [ ] No [ ]
*PRX30D	10/1GBASE-PRX-D3 or 10/1GBASE-PRX-U3 PMD	75.4, 75.5	Maximum channel insertion loss of 29 dB	O/1	Yes [ ] No [ ]
*INS	Installation/Cable	75.4.1	Items marked with INS include installation practices and cable specifications not applicable to PHY manufacturer	O	Yes [ ] No [ ]

75.10.4 PICS Proforma Tables for Physical Medium Dependent (PMD) Sublayer and Medium for Passive Optical Networks, Type 10GBASE–PR and 10/1GBASE–PRX

75.10.4.1 PMD Functional Specifications

Item	Feature	Subclause	Value/Comment	Status	Support
FN1	Transmit function	75.3.3	Conveys bits from PMD service interface to MDI	M	Yes [ ]
FN2	Transmitter optical signal	75.3.3	Higher optical power transmitted is a logic 1	M	Yes [ ]
FN3	Receive function	75.3.4	Conveys bits from MDI to PMD service interface	M	Yes [ ]
FN4	Receiver optical signal	75.3.4	Higher optical power received is a logic 1	M	Yes [ ]
FN5	ONU signal detect function	75.3.5.1	Mapping to PMD service interface	M	Yes [ ]
FN6	ONU signal detect parameter	75.3.5.1	Generated according to Table 75.4	M	Yes [ ]
FN7	OLT signal detect function	75.3.5.2	Mapping to PMD service interface	O/2	Yes [ ]
FN8	OLT signal detect function	75.3.5.2	Provided by higher layer	O/2	Yes [ ]
FN9	OLT signal detect parameter	75.3.5.1	Generated according Table 75.4	O	Yes [ ]

75.10.4.2 PMD to MDI Optical Specifications for 10GBASE–PR–D1

Item	Feature	Subclause	Value/Comment	Status	Support
PRD1F1	10GBASE-PR-D1 transmitter	75.4.1	Meets specifications in Table 75.5	PRD1F1:M	Yes [ ] N/A [ ]
PRD1F2	10GBASE-PR-D1 receiver	75.4.2	Meets specifications in Table 75.6	PRD1F2:M	Yes [ ] N/A [ ]
PRD1F3	10GBASE-PR-D1 stressed receiver sensitivity	75.4.2	Meets specifications in Table 75.6	PRD1F3:O	Yes [ ] No [ ] N/A [ ]
PRD1F4	10GBASE-PR-D1 receiver damage threshold	75.4.2	If the receiver does not meet the damage requirements in Table 75.6 then label accordingly	PRD1F4:M	Yes [ ] N/A [ ]

75.10.4.3 PMD to MDI Optical Specifications for 10GBASE–PR–D2
- - - - - - - -

75.10.4.4 PMD to MDI Optical Specifications for 10GBASE–PR–D3
- - - - - - - -

75.10.4.5 PMD to MDI Optical Specifications for 10/1GBASE–PRX–D1
- - - - - - - -

75.10.4.6 PMD to MDI Optical Specifications for 10/1GBASE–PRX–D2
- - - - - - - -

75.10.4.7 PMD to MDI Optical Specifications for 10/1GBASE–PRX–D3
- - - - - - - -

75.10.4.8 PMD to MDI Optical Specifications for 10GBASE–PR–U1
- - - - - - - -

75.10.4.9 PMD to MDI Optical Specifications for 10GBASE–PR–U3
- - - - - - - -

75.10.4.10 PMD to MDI Optical Specifications for 10/1GBASE–PRX–U1
- - - - - - - -

75.10.4.11 PMD to MDI Optical Specifications for 10/1GBASE–PRX–U2
- - - - - - - -

75.10.4.12 PMD to MDI Optical Specifications for 10/1GBASE–PRX–U3
- - - - - - - -

75.10.4.13 Definitions of Optical Parameters and Measurement Methods
- - - - - - - -

77. MULTIPOINT MAC CONTROL FOR 10G   EPON

77.1   Overview

This clause deals with the mechanism and control protocols required in order to reconcile the 10 Gb/s P2MP topology into the Ethernet framework. The P2MP medium is a passive optical network (PON), an optical network with no active elements in the signal's path from source to destination. The only interior elements used in a PON are passive optical components, such as optical fiber, splices, and splitters. When combined with the Ethernet protocol, such a network is referred to as Ethernet passive optical network (EPON).

P2MP is an asymmetric medium based on a tree (or tree-and-branch) topology. The DTE connected to the trunk of the tree is called optical line terminal (OLT) and the DTEs connected at the branches of the tree are called optical network units (ONU). The OLT typically resides at the service provider's facility, while the ONUs are located at the subscriber premises.

In the downstream direction (from the OLT to an ONU), signals transmitted by the OLT pass through a 1:N passive splitter (or cascade of splitters) and reach each ONU. In the upstream direction (from the ONUs to the OLT), the signal transmitted by an ONU would only reach the OLT, but not other ONUs. To avoid data collisions and increase the efficiency of the subscriber access network, the ONU's transmissions are arbitrated. This arbitration is achieved by allocating a transmission window (grant) to each ONU. An ONU defers transmission until its grant arrives. When the grant arrives, the ONU transmits frames at wire speed during its assigned time slot.

A simplified P2MP topology example is depicted in Figure 77.1. Clause 67 provides additional examples of P2MP topologies.

Topics dealt with in this clause include allocation of upstream transmission resources to different ONUs, discovery and registration of ONUs

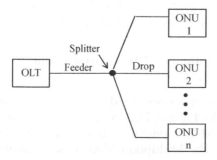

Figure 77.1   PON topology example.

into the network, and reporting of congestion to higher layers to allow for dynamic bandwidth allocation schemes and statistical multiplexing across the PON.

This clause does not deal with topics including bandwidth allocation strategies, authentication of end-devices, quality-of-service definition, provisioning, or management.

This clause specifies the multipoint control protocol (MPCP) to operate an optical multipoint network by defining a Multipoint MAC Control sublayer as an extension of the MAC Control sublayer defined in Clause 31, and supporting current and future operations as defined in Clause 31 and annexes.

Each PON consists of a node located at the root of the tree assuming the role of OLT, and multiple nodes located at the tree leaves assuming roles of ONUs. The network operates by allowing only a single ONU to transmit in the upstream direction at a time. The MPCP located at the OLT is responsible for timing the different transmissions. Reporting of congestion by the different ONUs may assist in optimally allocating the bandwidth across the PON.

Automatic discovery of end stations is performed, culminating in registration through binding of an ONU to an OLT port by allocation of a Logical Link ID (see LLID in 76.2.6.1.3.2), and dynamic binding to a MAC connected to the OLT.

77.1.1 Goals and Objectives

The goals and objectives of this clause are the definition of a point-to-multipoint Ethernet network utilizing an optical medium.

Specific objectives met include the following:

a. Support of point-to-point Emulation (P2PE) as specified
b. Support multiple LLIDs and MAC Clients at the OLT
c. Support a single LLID per ONU
d. Support a mechanism for single copy broadcast
e. Flexible architecture allowing dynamic allocation of bandwidth
f. Use of 32 bit time stamp for timing distribution
g. MAC Control based architecture
h. Ranging of discovered devices for improved network performance
i. Continuous ranging for compensating round trip time variation

77.1.2 Position of Multipoint MAC Control within the IEEE 802.3 Hierarchy

Multipoint MAC Control defines the MAC control operation for optical point-to-multipoint networks. Figures 77.2 and 77.3 depict the architectural positioning of the Multipoint MAC Control sublayer with respect to the MAC and the MAC Control client. The Multipoint MAC Control sublayer takes the place

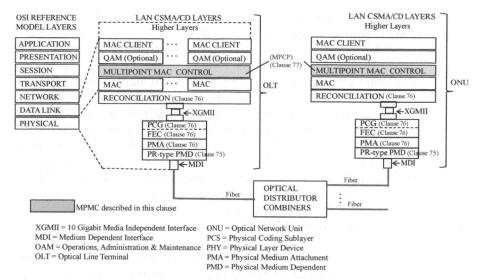

Figure 77.2 Relationship of Multipoint MAC Control and the OSI protocol stack for 10/10G–EPON (10 GB/s downstream and 10 Gb/s upstream).

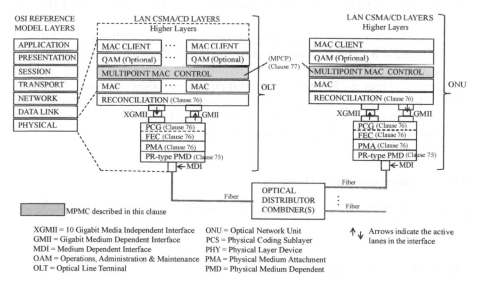

Figure 77.3 Relationship of Multipoint MAC Control and the OSI protocol stack for 10/1G–EPON (10 Gb/s downstream and 1 Gb/s upstream).

of the MAC Control sublayer to extend it to support multiple clients and additional MAC control functionality.

Multipoint MAC Control is defined using the mechanisms and precedents of the MAC Control sublayer. The MAC Control sublayer has extensive functionality designed to manage the real-time control and manipulation of MAC sublayer operation. This clause specifies the extension of the MAC Control mechanism to manipulate multiple underlying MACs simultaneously. This clause also specifies a specific protocol implementation for MAC Control.

The Multipoint MAC Control sublayer is specified such that it can support new functions to be implemented and added to this standard in the future. MultiPoint Control Protocol (MPCP), the management protocol for P2MP is one of these protocols. Non-real-time, or quasi-static control (e.g., configuration of MAC operational parameters) is provided by Layer Management. Operation of the Multipoint MAC Control sublayer is transparent to the MAC.

As depicted in Figures 77.2 and 77.3, the layered system instantiates multiple MAC entities, using a single Physical Layer. The individual MAC instances offer a point-to-point emulation service between the OLT and the ONU. An additional MAC is instantiated to communicate to all 10G–EPON ONUs at once. This instance takes maximum advantage of the broadcast nature of the downstream channel by sending a single copy of a frame that is received by all 10G–EPON ONUs. This MAC instance is referred to as Single Copy Broadcast (SCB).

The ONU only requires one MAC instance since frame filtering operations are done at the RS layer before reaching the MAC. Therefore, MAC and layers above are emulation-agnostic at the ONU (see 76.2.6.1.3).

Although Figures 77.2 and 77.3 and supporting text describe multiple MACs within the OLT, a single unicast MAC address may be used by the OLT. Within the EPON Network, MACs are uniquely identified by their LLIDs, which are dynamically assigned by the registration process.

77.1.3 Functional Block Diagram

Figure 77.4 provides a functional block diagram of the Multipoint MAC Control architecture.

77.1.4 Service Interfaces

The MAC Client communicates with the Control Multiplexer using the standard service interface specified in 2.3. Multipoint MAC Control communicates with the underlying MAC sublayer using the standard service interface specified in Annex 4A.3.2. Similarly, Multipoint MAC Control communicates internally using primitives and interfaces consistent with definitions in Clause 31.

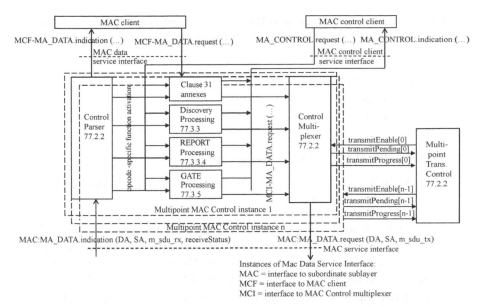

Figure 77.4 Multipoint MAC Control functional block diagram.

77.1.5 State Diagram Conventions

The body of this standard comprises state diagrams, including the associated definitions of variables, constants, and functions. Should there be a discrepancy between a state diagram and descriptive text, the state diagram prevails.

The notation used in the state diagrams follows the conventions of 21.5. State diagram timers follow the conventions of 14.2.3.2 augmented as follows:

a. (start x_timer, y) sets expiration of y to timer x_timer.
b. (stop x_timer) aborts the timer operation for x_timer asserting x timer_ not_done indefinitely.

The state diagrams use an abbreviation **MACR** as a shorthand form for **MA_CONTROL**.request and **MACI** as a shorthand form for **MA_CONTROL**.indication.

The vector notations used in the state diagrams for bit vector use 0 to mark the first received bit and so on (e.g., data[0:15]), following the conventions of 3.1 for bit ordering. When referring to an octet vector, 0 is used to mark the first received octet and so on (e.g., m_sdu[0..1]).

a < b: A function that is used to compare two (cyclic) time values. Returned value is true when b is larger than a allowing for wrap around of a and b. The comparison is made by subtracting b from a and testing the MSB. When

MSB(a–b) = 1 the value true is returned, else false is returned. In addition, the following functions are defined in terms of a < b (where ! means complement of):

a > b is equivalent to !(a < b or a = b)

a ≥ b is equivalent to !(a < b)

a ≤ b is equivalent to !(a > b)

77.2 Multipoint MAC Control Operation

As depicted in Figure 77.4, the Multipoint MAC Control functional block comprises the following functions:

   a. Multipoint Transmission Control. This block is responsible for synchronizing Multipoint MAC Control instances associated with the Multipoint MAC Control. This block maintains the Multipoint MAC Control state and controls the multiplexing functions of the instantiated MACs.
   b. Multipoint MAC Control Instance n. This block is instantiated for each MAC and respective MAC and MAC Control clients associated with the Multipoint MAC Control. It holds all the variables and state associated with operating all MAC Control protocols for the instance.
   c. Control Parser. This block is responsible for parsing MAC Control frames, and interfacing with Clause 31 entities, the opcode specific blocks, and the MAC Client.
   d. Control Multiplexer. This block is responsible for selecting the source of the forwarded frames.
   e. Clause 31 annexes. This block holds MAC Control actions as defined in Clause 31 annexes for support of legacy and future services.
   f. Discovery, Report, and Gate Processing. These blocks are responsible for handling the MPCP in the context of the MAC.

77.2.1 Principles of Multipoint MAC Control

As depicted in Figure 77.4, Multipoint MAC Control sublayer may instantiate multiple Multipoint MAC Control instances in order to interface multiple MAC and MAC Control clients above with multiple MACs below. A unique unicast MAC instance is used at the OLT to communicate with each ONU. The individual MAC instances utilize the point-to-point emulation service between the OLT and the ONU as defined in 76.2.

At the ONU, a single MAC instance is used to communicate with a MAC instance at the OLT. In that case, the Multipoint MAC Control contains only a single instance of the Control Parser/Multiplexer function.

Multipoint MAC Control protocol supports several MAC and client interfaces. Only a single MAC interface and Client interface is enabled for transmission at a time. There is a tight mapping between a MAC service interface and a Client service interface. In particular, the assertion of the MAC:MA_DATA.indication primitive in MAC j leads to the assertion of the MCF:MA_DATA.indication primitive to Client j. Conversely, the assertion of the request service interface in Client i leads to the assertion of the MAC:MA_DATA. request primitive of MAC i. Note that the Multipoint MAC sublayer need not receive and transmit packets associated with the same interface at the same time. Thus the Multipoint MAC Control acts like multiple MAC Controls bound together with common elements.

The scheduling algorithm is implementation dependent, and is not specified for the case where multiple transmit requests happen at the same time.

The reception operation is as follows. The Multipoint MAC Control instances generate MAC:MA_DATA.indication service primitives continuously to the underlying MAC instances. Since these MACs are receiving frames from a single PHY only one frame is passed from the MAC instances to Multipoint MAC Control. The MAC instance responding to the MAC:MA_DATA.indication is referred to as the enabled MAC, and its service interface is referred to as the enabled MAC interface. The MAC passes to the Multipoint MAC Control sublayer all valid frames. Invalid frames, as specified in 3.4, are not passed to the Multipoint MAC Control sublayer in response to a MAC:MA_DATA.indication service primitive.

The enabling of a transmit service interface is performed by the Multipoint MAC Control instance in collaboration with the Multipoint Transmission Control. Frames generated in the MAC Control are given priority over MAC Client frames, in effect, prioritizing the MA_CONTROL primitive over the MCF:MA_DATA primitive, and for this purpose MCF:MA_DATA.request primitives may be delayed, discarded or modified in order to perform the requested MAC Control function. For the transmission of this frame, the Multipoint MAC Control instance enables forwarding by the MAC Control functions, but the MAC Client interface is not enabled. The reception of a frame in a MAC results in generation of the MAC:MA_DATA.indication primitive on that MAC's interface. Only one receive MAC interface is enabled at any given time since there is only one PHY interface.

The information of the enabled interfaces is stored in the controller state variables, and accessed by the Multiplexing Control block.

The Multipoint MAC Control sublayer uses the services of the underlying MAC sublayer to exchange both data and control frames.

Receive operation (MAC:MA_DATA.indication) at each instance:

a. A frame is received from the underlying MAC
b. The frame is parsed according to Length/Type field
c. MAC Control frames are demultiplexed according to opcode and forwarded to the relevant processing functions

d. Data frames (see 31.5.1) are forwarded to the MAC Client by asserting MCF:MA_DATA.indication primitives Transmit operation (MAC:MA_DATA.request) at each instance:

e. The MAC Client signals a frame transmission by asserting MCF:MA_DATA.request, or

f. A protocol processing block attempts to issue a frame, as a result of a previous MA_CONTROL.request or as a result of an MPCP event that generates a frame.

g. When allowed to transmit by the Multipoint Transmission Control block, the frame is forwarded.

77.2.1.1 Ranging and Timing Process (Figure 77.5)

77.2.1.1   Ranging and Timing Process (Figure 77.5)   Both the OLT and the ONU have 32 bit counters that increment every 16 ns. These counters provide a local time stamp. When either device transmits an MPCPDU, it maps its counter value into the timestamp field. The time of transmission of the first octet of the MPCPDU frame from the MAC Control to the MAC is taken as the reference time used for setting the timestamp value.

When the ONU receives MPCPDUs, it sets its counter according to the value in the timestamp field in the received MPCPDU.

When the OLT receives MPCPDUs, it uses the received timestamp value to calculate or verify a round trip time between the OLT and the ONU. The Round Trip Time (RTT) is equal to the difference between the timer value and the value in the timestamp field. The calculated RTT is notified to the client via the MA_CONTROL.indication primitive. The client can use this RTT for the ranging process.

A condition of *timestamp drift error* occurs when the difference between OLT's and ONU's clocks exceeds some predefined threshold. This condition can be independently detected by the OLT or an ONU. The OLT detects this condition when an absolute difference between new and old RTT values measured for a given ONU exceeds the value of guardThresholdOLT (see 77.2.2.1), as shown in Figure 77.11. An ONU detects the timestamp drift error condition when absolute difference between a time stamp received in an MPCPDU and the localTime counter exceeds guardThresholdONU (see 77.2.2.1), as is shown in Figure 77.12.

77.2.2   Multipoint transmission control, Control Parser, and Control Multiplexer

The purpose of the multipoint transmission control is to allow only one of the multiple MAC clients to transmit to its associated MAC and subsequently to the RS layer at one time by only asserting one transmitEnable signal at a time. (Figure 77.6)

Multipoint MAC Control Instance n function block communicates with the Multipoint Transmission Control using transmitEnable[n], transmitPending[n],

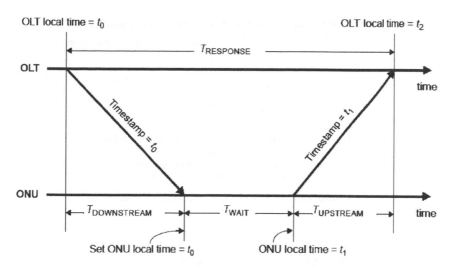

$T_{DOWNSTREAM}$ = downstream propagation delay

$T_{UPSTREAM}$ = upstream propagation delay

T_{WAIT} = wait time at ONU = $t_1 - t_0$

$T_{RESPONSE}$ = response time at OLT = $t_2 - t_0$

$$RTT = T_{DOWNSTREAM} + T_{UPSTREAM} = T_{RESPONSE} - T_{WAIT} = (t_2 - t_0) - (t_1 - t_0) = t_2 - t_1$$

Figure 77.5 Round trip time calculation.

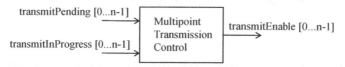

Figure 77.6 Multipoint Transmission Control service interfaces.

and transmitInProgress[n] state variables (see Figure 77.4). The Control Parser is responsible for opcode independent parsing of MAC frames in the reception path. By identifying MAC Control frames, demultiplexing into multiple entities for event handling is possible. Interfaces are provided to existing Clause 31 entities, functional blocks associated with MPCP, and the MAC Client.

The Control Multiplexer is responsible for forwarding frames from the MAC Control opcode-specific functions and the MAC Client to the MAC. Multiplexing is performed in the transmission direction. Given multiple MCF:MA_DATA.request primitives from the MAC Client, and MA_

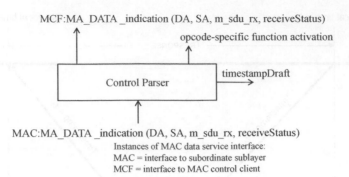

Figure 77.7 Control Parser service interfaces.

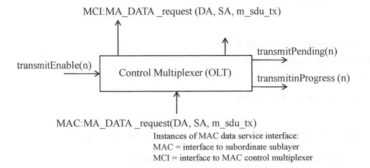

NOTE-MAC:MA-DATA_request primitive may be issued from multiple MAC Control processing blocks

Figure 77.8 OLT Control Multiplexer service interfaces.

CONTROL.request primitives from the MAC Control Clients, a single MAC:MA_DATA.request service primitive is generated\ for transmission. At the OLT, multiple MAC instances share the same Multipoint MAC Control, as a result, the transmit block is enabled based on an external control signal housed in Multipoint Transmission Control for transmission overlap avoidance. At the ONU, the Gate Processing functional block interfaces for upstream transmission administration. (Figures 77.7–77.9)

77.2.2.1 Constants
- - - - - - - -

77.2.2.2 Counters
- - - - - - - -

77.2.2.3 Variables
- - - - - - - -

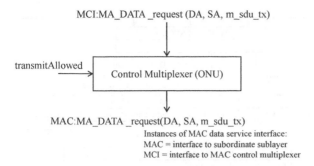

MCI:MA_DATA _request (DA, SA, m_sdu_tx)

transmitAllowed

Control Multiplexer (ONU)

MAC:MA_DATA _request(DA, SA, m_sdu_tx)
Instances of MAC data service interface:
MAC = interface to subordinate sublayer
MCI = interface to MAC control multiplexer

NOTE-MAC:MA-DATA_request primitive may be issued from multiple MAC Control processing blocks

Figure 77.9   ONU Control Multiplexer service interfaces.

77.2.2.4   *Functions*

- - - - - - - -

77.2.2.5   *Timers*

- - - - - - - -

77.2.2.6   *Messages*

- - - - - - - -

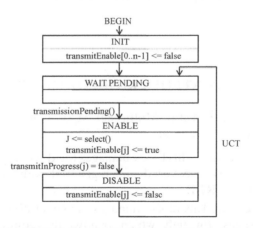

BEGIN

INIT
transmitEnable[0..n-1] <= false

WAIT PENDING

transmissionPending()

ENABLE
J <= select()
transmitEnable[j] <= true

UCT

transmitInProgress(j) = false

DISABLE
transmitEnable[j] <= false

Figure 77.10   OLT Multipoint Transmission Control state diagram.

77.2.2.7   *State Diagrams*   The Multipoint transmission control function in the OLT shall implement state diagram shown in Figure 77.10. Control parser function in the OLT shall implement state diagram shown in Figure 77.11. Control parser function in the ONU shall implement state diagram shown in Figure 77.12. Control multiplexer function in the OLT shall implement state diagram shown in Figure 77.13. Control multiplexer function in the ONU shall implement state diagram shown in Figure 77.14.

(Figures 77.12–77.14 not reproduced here.)

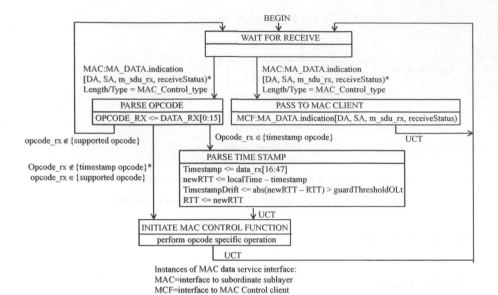

Figure 77.11   OLT Control Parser state diagram. NOTE—The opcode-specific operation is launched as a parallel process by the MAC Control sublayer, and not as a synchronous function. Progress of the generic MAC Control Receive state diagram (as shown in this figure) is not implicitly impeded by the launching of the opcode specific function.

- - - - - - - -

77.3   MULTIPOINT CONTROL PROTOCOL (MPCP)

As depicted in Figure 77.4, the Multipoint MAC Control functional block comprises the following functions:

a. Discovery Processing. This block manages the discovery process, through which an ONU is discovered and registered with the network while compensating for RTT.

b. Report Processing. This block manages the generation and collection of report messages, through which bandwidth requirements are sent upstream from the ONU to the OLT.

c. Gate Processing. This block manages the generation and collection of gate messages, through which multiplexing of multiple transmitters is achieved.

As depicted in Figure 77.4, the layered system may instantiate multiple MAC entities, using a single Physical Layer. Each instantiated MAC communicates with an instance of the opcode specific functional blocks through the Multipoint MAC Control. In addition some global variables are shared across the multiple instances. Common state control is used to synchronize the multiple MACs using MPCP procedures. Operation of the common state control is generally considered outside the scope of this document.

77.3.1 Principles of Multipoint Control Protocol

Multipoint MAC Control enables a MAC Client to participate in a point-to-multipoint optical network by allowing it to transmit and receive frames as if it was connected to a dedicated link. In doing so, it employs the following principles and concepts:

a. A MAC client transmits and receives frames through the Multipoint MAC Control sublayer.

b. The Multipoint MAC Control decides when to allow a frame to be transmitted using the client interface Control Multiplexer.

c. Given a transmission opportunity, the MAC Control may generate control frames that would be transmitted in advance of the MAC Client's frames, utilizing the inherent ability to provide higher priority transmission of MAC Control frames over MAC Client frames.

d. Multiple MACs operate on a shared medium by allowing only a single MAC to transmit upstream at any given time across the network using a time-division multiple access (TDMA) method.

e. Such gating of transmission is orchestrated through the Gate Processing function.

f. New devices are discovered in the network and allowed transmission through the Discovery Processing function.

g. Fine control of the network bandwidth distribution can be achieved using feedback mechanisms supported in the Report Processing function.

h. The operation of P2MP network is asymmetric, with the OLT assuming the role of master, and the ONU assuming the role of slave.

77.3.2 Compatibility Considerations

77.3.2.1 PAUSE Operation Even though MPCP is compatible with flow control, optional use of flow control may not be efficient in the case of large propagation delay. If flow control is implemented, then the timing constraints in Clause 31B supplement the constraints found at 77.3.2.4.

NOTE—MAC at an ONU can receive frames from unicast channel and SCB channel. If the SCB channel is used to broadcast data frames to multiple ONUs, the ONU's MAC may continue receiving data frames from SCB channel even after the ONU has issued a PAUSE request to its unicast remote-end.

77.3.2.2 Optional Shared LAN Emulation   By combining P2PE, suitable filtering rules at the ONU, and suitable filtering and forwarding rules at the OLT, it is possible to emulate an efficient shared LAN. Support for shared LAN emulation is optional, and requires an additional layer above the MAC, which is out of scope for this document. Thus, shared LAN emulation is introduced here for informational purposes only.

Specific behaviour of the filtering layer at the RS is specified in 76.2.6.1.3.2.

77.3.2.3 Multicast and Single Copy Broadcast Support   In the downstream direction, the PON is a broadcast medium. In order to make use of this capability for forwarding broadcast frames from the OLT to multiple recipients without frame duplication for each ONU, the SCB support is introduced.

The OLT has at least one MAC associated with every ONU. In addition one more MAC at the OLT is marked as the SCB MAC. The SCB MAC handles all downstream broadcast traffic, but is never used in the upstream direction for client traffic, except for client registration. Optional higher layers may be implemented to perform selective broadcast of frames. Such layers may require additional MACs (multicast MACs) to be instantiated in the OLT for some or all ONUs increasing the total number of MACs beyond the number of ONUs + 1.

When connecting the SCB MAC to an IEEE 802.1D bridge port it is possible that loops may be formed due to the broadcast nature. Thus it is recommended that this MAC not be connected to an IEEE 802.1D bridge port.

Configuration of SCB channels as well as filtering and marking of frames for support of SCB is defined in 76.2.6.1.3.2 for 10G–EPON compliant Reconciliation Sublayers.

77.3.2.4 Delay Requirements   The MPCP protocol relies on strict timing based on distribution of time stamps. A compliant implementation needs to guarantee a constant delay through the MAC and PHY in order to maintain the correctness of the timestamping mechanism. The actual delay is implementation dependent; however, a complying implementation shall maintain a delay variation of no more than 1 time_quantum through the MAC.

The OLT shall not grant less than 1024 time_quanta into the future, in order to allow the ONU processing time when it receives a gate message. The ONU shall process all messages in less than this period. The OLT shall not issue more than one message every 1024 time_quanta to a single ONU. The unit of time_quantum is defined in 77.2.2.1.

77.3.3 Discovery Processing

Discovery is the process whereby newly connected or off-line ONUs are provided access to the PON. The process is driven by the OLT, which periodically makes available Discovery Windows during which off-line ONUs are given the opportunity to make themselves known to the OLT. The periodicity of these windows is unspecified and left up to the implementor. The OLT signifies that a discovery period is occurring by broadcasting a discovery GATE MPCPDU, which includes the starting time and length of the discovery window, along with the Discovery Information flag field, as defined in 77.3.6.1. With the appropriate settings of individual flags contained in this 16 bit wide field, the OLT notifies all the ONUs about its upstream and downstream channel transmission capabilities. Note that the OLT may simultaneously support more than one data rate in the given transmission direction.

Off-line ONUs, upon receiving a Discovery GATE MPCPDU, wait for the period to begin and then transmit a REGISTER_REQ MPCPDU to the OLT. Discovery windows are unique in that they are the only times when multiple ONUs can access the PON simultaneously, and transmission overlap can occur. In order to reduce transmission overlaps, a contention algorithm is used by all ONUs. Measures are taken to reduce the probability for overlaps by artificially simulating a random distribution of distances from the OLT. Each ONU waits a random amount of time before transmitting the REGISTER_REQ MPCPDU that is shorter than the length of the discovery window. It should be noted that multiple valid REGISTER_REQ MPCPDUs can be received by the OLT during a single discovery window. Included in the REGISTER REQ MPCPDU is the ONU's MAC address and number of maximum pending grants. Additionally, a registering ONU notifies the OLT of its transmission capabilities in the upstream and downstream channels by setting appropriately the flags in the Discovery Information field, as specified in 77.3.6.3.

Note that even though a compliant ONU is not prohibited from supporting more than one data rate in any transmission channel, it is expected that a single supported data rate for upstream and downstream channel is indicated in the Discovery Information field. Moreover, in order to assure maximum utilization of the upstream channel and to decrease the required size of the guard band between individual data bursts, the registering ONU notifies the OLT of the laser on/off times, by setting appropriate values in the Laser On Time and Laser Off Time fields, where both values are expressed in the units of time_quanta.

Upon receipt of a valid REGISTER_REQ MPCPDU, the OLT registers the ONU, allocating and assigning a new port identity (LLID), and bonding a corresponding MAC to the LLID.

The next step in the process is for the OLT to transmit a REGISTER MPCPDU to the newly discovered ONU, which contains the ONU's LLID, and the OLT's required synchronization time. Moreover, the OLT echoes the maximum number of pending grants. The OLT also sends the target value of laser on time and laser off time, which may be different than laser on time and laser off time delivered by the ONU in the REGISTER_REQ MPCPDU.

The OLT now has enough information to schedule the ONU for access to the PON and transmits a standard GATE message allowing the ONU to

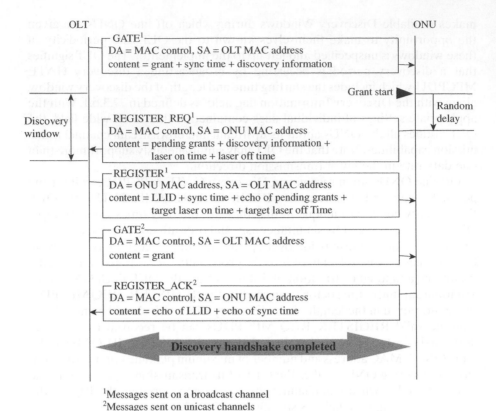

OLT   ONU

┌─ GATE[1]
│ DA = MAC control, SA = OLT MAC address
│ content = grant + sync time + discovery information

Grant start ▶

Random delay

Discovery window

┌─ REGISTER_REQ[1]
│ DA = MAC control, SA = ONU MAC address
│ content = pending grants + discovery information +
│   laser on time + laser off time

┌─ REGISTER[1]
│ DA = ONU MAC address, SA = OLT MAC address
│ content = LLID + sync time + echo of pending grants +
│   target laser on time + target laser off Time

┌─ GATE[2]
│ DA = MAC control, SA = OLT MAC address
│ content = grant

┌─ REGISTER_ACK[2]
│ DA = MAC control, SA = ONU MAC address
│ content = echo of LLID + echo of sync time

Discovery handshake completed

[1]Messages sent on a broadcast channel
[2]Messages sent on unicast channels

Figure 77.15   Discovery handshake message exchange.

transmit a REGISTER_ACK. Upon receipt of the REGISTER_ACK, the discovery process for that ONU is complete, the ONU is registered and normal message traffic can begin. It is the responsibility of Layer Management to perform the MAC bonding, and start transmission from/to the newly registered ONU. The discovery message exchange is illustrated in Figure 77.15.

There may exist situations when the OLT requires that an ONU go through the discovery sequence again and reregister. Similarly, there may be situations where an ONU needs to inform the OLT of its desire to deregister. The ONU can then reregister by going through the discovery sequence. For the OLT, the REGISTER message may indicate a value, Reregister or Deregister, that if either is specified forces the receiving ONU into reregistering. For the ONU, the REGISTER_REQ message contains the Deregister bit that signifies to the OLT that this ONU should be deregistered. (Figures 77.16–77.18).

(Figures 77.16–77.18 not reproduced here.)

77.3.3.1 Constants

- - - - - - - -

77.3.3.2 Variables

- - - - - - - -

77.3.3.3 Functions

- - - - - - - -

77.3.3.4 Timers

- - - - - - - -

77.3.3.5 Messages

- - - - - - - -

77.3.3.6 State Diagrams The Discovery Process in the OLT shall implement the discovery window setup state diagram shown in Figure 77.19, request processing state diagram as shown in Figure 77.20, register processing state diagram as shown in Figure 77.21, and final registration state diagram as shown in Figure 77.22. The discovery process in the ONU shall implement the registration state diagram as shown in Figure 77.23.

Instantiation of state diagrams as described in Figures 77.19–77.21 is performed only at the Multipoint MAC Control instances attached to the broadcast LLID (0x7FFE). Instantiation of state diagrams as described in Figures 77.22 and 77.3 is performed for every Multipoint MAC Control instance, except the instance attached to the broadcast channel.

(Figures 77.19–77.23 not reproduced here.)

- - - - - - - -

- - - - - - - -

INDEX

The ComSoc Guide to Passive Optical Networks: Enhancing the Last Mile Access,
First Edition. Stephen Weinstein, Yuanqiu Luo, Ting Wang.
© 2012 Institute of Electrical and Electronics Engineers. Published 2012 by John Wiley & Sons, Inc.

Printed and bound by CPI Group (UK) Ltd, Croydon, CR0 4YY

27/10/2024

14580257-0001